Editor
Jamie Wu Liu, M.A.

Editorial Project Manager
Elizabeth Morris, Ph.D.

Editor-in-Chief
Sharon Coan, M.S. Ed.

Creative Director
Elayne Roberts

Product Manager
Phil Garcia

Imaging
Alfred Lau

Acknowledgements
Word® software is ©1983–2000 Microsoft Corporation. All rights reserved. *Word* is a registered trademark of Microsoft Corporation.

Microsoft® Word™: Simple Projects

Grades 4-6

Publisher
Mary D. Smith, M.S. Ed.

Author

Mindy Pines

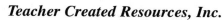

Teacher Created Resources

Teacher Created Resources, Inc.
6421 Industry Way
Westminster, CA 92683
www.teachercreated.com
ISBN: 978-1-57690-728-3
©2000 Teacher Created Resources, Inc.
Reprinted, 2010
Made in U.S.A.

The classroom teacher may reproduce copies of materials in this book for classroom use only. The reproduction of any part for an entire school or school system is strictly prohibited. No part of this publication may be transmitted, stored, or recorded in any form without written permission from the publisher.

Table of Contents

Introduction

Any experienced teacher can tell you that trends in education, in which we're all expected to jump onto the bandwagon, are common. Often, these trends have nothing to do with what really works well with students. Sometimes they do. When computers first became the "thing," I was cynical. After all, I'd been through the rejection of phonics and its replacement by whole language, only to return to phonics, this time more sanely as a part of an overall program. I'd also seen the replacement of computation drills by calculator-assisted problem solving only to return to computation drills as part of an overall math program.

At first, I resented the hype about technology. It angered me to see so much money spent on computers and software that could be used only by a few students at a time when there weren't enough books for each student in my classroom. I resisted computers. However, I soon discovered how word processing made my life easier, saving hundreds of hours of rewrites. As a teacher, I appreciated how much easier it was to keep student records and how I could save my lesson plans. Not only could I revise and reuse my lessons year after year, but if you ever saw how messy my handwriting is, not to mention how disorganized my desk and file cabinets are, you'd understand how much I've benefited from learning to use computers.

Not one to jump onto bandwagons though, I spent twice as much energy resisting using computers with my students in the classroom than figuring out ways to use them until I had a particularly difficult class with some students who had histories of delinquency, truancy and failing grades. They weren't the type of students who completed their assignments because they wanted to please their teachers. But when I took these students to the computer lab or let them use the single computer in back of my classroom, I noticed that they stayed on task, and on their chairs, and when they talked to one another, it was about the work. If there was anything I could use to make my job easier, I was going to use it. That's when I started really thinking about how I could get the most from using computers in the classroom. What I discovered holds true for all classes I've had since, not just with at-risk students who may have behavior problems, but with motivated, average, and gifted students as well. I didn't even have to change what I taught or how I taught it. Students take to computers naturally. They enjoy working at them. Accordingly, computers make teachers' jobs easier, the students' jobs more fun and the atmosphere more positive. At the same time the students are gaining computer skills which today are necessary to secure practically every kind of job.

The activities in this book are designed to support, not change, what you already do. They are appropriate for students both in the computer lab or at a single computer in the classroom. They support the skills you are already teaching while students learn how to use some of the many features of *Microsoft Word*. They are designed to be merely a part of a well-rounded, well-planned curriculum.

Simple Projects for Microsoft Word includes curriculum-centered activities in four major curricular areas: language arts, social studies, science, and math. Each lesson is assigned to a particular area, but very easily adapted to other areas. Most lessons have templates provided on the CD-ROM for use with *Microsoft Word 97 for Windows* and *Word 98 for Macintosh*. Though *Simple Projects for Microsoft Word* was written mostly for the latest version of *Word*, many of the same tools, features, and instructions apply to previous versions of *Word*. In the "**Appendix**", we offer tips on how to save documents for use in different versions of *Word*. We also offer tips on using many features, specifically helpful to teachers, about which you may be unaware. *Simple Projects for Microsoft Word* was designed to make your job easier and your students' jobs more fun. Enjoy!

Using this Book

The **Toolbar** illustrations below should help you locate the tools and buttons you'll need to follow the instructions given in *Simple Projects for Microsoft Word*. For more detailed instructions on how to use *Microsoft Word*, see *Microsoft Word for Terrified Teachers* by Paula G. Patton and Karla Neeley Hase, which was also published by Teacher Created Resources.

The top of your screen should appear similar to the illustration below. There may be some variance, depending upon your version of *Microsoft Word*, computer platform, and operating system.

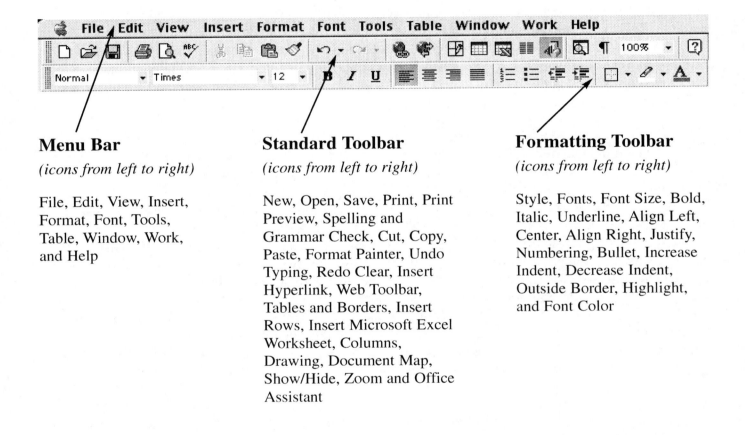

Menu Bar
(icons from left to right)

File, Edit, View, Insert, Format, Font, Tools, Table, Window, Work, and Help

Standard Toolbar
(icons from left to right)

New, Open, Save, Print, Print Preview, Spelling and Grammar Check, Cut, Copy, Paste, Format Painter, Undo Typing, Redo Clear, Insert Hyperlink, Web Toolbar, Tables and Borders, Insert Rows, Insert Microsoft Excel Worksheet, Columns, Drawing, Document Map, Show/Hide, Zoom and Office Assistant

Formatting Toolbar
(icons from left to right)

Style, Fonts, Font Size, Bold, Italic, Underline, Align Left, Center, Align Right, Justify, Numbering, Bullet, Increase Indent, Decrease Indent, Outside Border, Highlight, and Font Color

Drawing Toolbar

Icons from left to right: Draw, Select Objects, Free Rotate, AutoShapes; Line, Arrow; Rectangle, Oval, Text Box, Insert WordArt, Fill Color, Line Color, Font Color, Line Style, Dash Style, Arrow Style, Shadow, and 3-D

Note: If the Drawing Toolbar is not already at the bottom of your screen, pull down the VIEW menu, select Toolbars, then Drawing from the drop-down menu.

Using this Book *(Cont.)*

About the CD-ROM

Turn to the back of this book and you will find a CD-ROM. It contains resource files for the student activities. The CD-ROM does not contain the *Microsoft Word* software application. *Microsoft Word* must be installed on your computer system prior to using this book and the resource CD-ROM.

There are two types of files on the CD-ROM. There are template files and sample files. Nearly every activity and lesson in *Microsoft Word Simple Projects* has associated CD-ROM files for use with *Microsoft Word 97 for Windows* and *Microsoft Word 98 for Macintosh*. You will find a CD-ROM **Filenames** index below.

The template files associated with the student lessons are already created for you. They are all on the CD-ROM. Template files are meant for you to open and immediately save under another name somewhere on your desktop, hard drive, or floppy diskette. That way the original template file is always intact and ready for you to use again and again. So, if you open a template file, before entering any text or data, click on the **FILE** menu and select **Save As**. Navigate to where you want to save your file, rename it, and then click on the **Save** button. Then you can begin entering text and data, knowing that the original file is still intact. That's all there is to using a template file!

Feel free to create a **Word Simple Projects** folder on your desktop or hard drive and download all the files from the CD-ROM into it for your convenience. Here's an organizational tip: Create a Student Folder named for each student in your class within that **Simple Projects** folder to keep the files they complete as they work through the book. Don't you just love being organized?

CD-ROM Filenames

Page	Activity	(Templates folder) Template Filename	(Samples folder) Sample Filename
6	Blue Ribbon Award	Ribbon.doc	Ribbons.doc
10	What a Character!	Character.doc	Characts.doc
16	You Don't Say!	Dialogue.doc	Says.doc
21	5-7-5		Haikus.doc
25	It's in the Mail!		Mails.doc
29	It's Negotiable	Negotiable.doc	Conflics.doc
33	Putting Things in Order	Timeline.doc	Times.doc
38	Culture Comparison	Culture.doc	Cultures.doc
42	Geography Word Shuffle	Shuffle.doc	Shuffles.doc
48	Floor Plan		Plans.doc
52	Campaign Promises	Campaign.doc	Camps.doc
56	Koalas, Kangaroos, and More	Animal.doc	Animals.doc
60	It's Classified Information	Sorting.doc	Sortings.doc
64	Testing, Testing:1,2,3		Testings.doc
68	What Goes Around	Cycle.doc	Cycles.doc
73	What's Next	Pattern.doc	Patterns.doc
77	Playful Polygons	Polygon.doc	Polygons.doc
81	All in a Day's Fun	FunDay.doc	FunDays.doc
85	It's All in How You Cut It	CutIt.doc	CutIts.doc
89	What's Your Angle?		Angles.doc

Blue Ribbon Award
Describing Positive Qualities

This Project

In this project your students will create an award certificate for a classmate to acknowledge his or her positive qualities. This project can also be adapted to acknowledge historical figures from social studies or a character in a literature book.

Note: If you use the template (*Ribbon*) provided on the CD-ROM, direct your students to follow the prompts to fill in the requested information.

Computer Skills

- word processing
- importing clip art

Before Beginning

- Students should talk about finding positive qualities in others. They should select a classmate about whom they will acknowledge some positive qualities.

- Students should complete the planning sheet on Page 8.

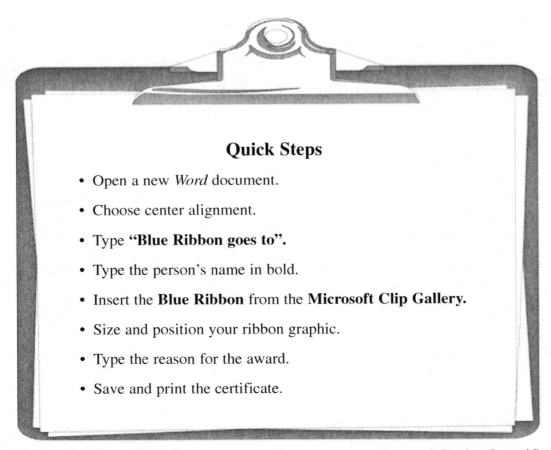

Quick Steps

- Open a new *Word* document.

- Choose center alignment.

- Type **"Blue Ribbon goes to"**.

- Type the person's name in bold.

- Insert the **Blue Ribbon** from the **Microsoft Clip Gallery.**

- Size and position your ribbon graphic.

- Type the reason for the award.

- Save and print the certificate.

Blue Ribbon Award
Describing Positive Qualities *(Cont.)*

Step 1   Open a new document.

Step 2   From the VIEW menu, choose **Toolbars**, then **Formatting** from the drop-down menu, if it is not already selected.

Step 3   Press the **Center** button.

Select a font and size (**26** to **36** points.)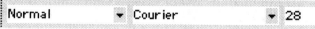

Type "**Blue Ribbon goes to**".

Press the **Enter** or **Return** key.

Step 4   Type the person's name for the award. Double-click on the name to highlight it.

Select **Bold** font style.   **B** *I* <u>U</u>

Step 5   Click at the end of the name. Press the **Enter** or **Return** key.

Step 6   From the INSERT menu, choose **Picture**, then **Clip Art** from the drop-down menu.

Step 7   In the **Microsoft Clip Gallery**, select **Entertainment** under the **Clip Art** tab. Choose the **Blue Ribbon** graphic. Click on the **Insert** button.

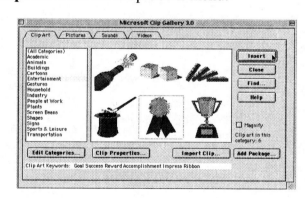

Step 8   Move the ribbon by clicking inside and dragging it to line up with the text on top. Size the blue ribbon by dragging the handles around the graphic.

Step 9   Click outside of the ribbon. Press the **Enter** or **Return** key a few times until the cursor appears under the ribbon.

Step 10 Click on the **Bold** button to deactivate the function. Type the word "**for**". Press the **Enter** or **Return** key.

Step 11 Type some positive qualities of this person that you want to acknowledge.

Step 12 Save and print your document.

Blue Ribbon Award
Describing Positive Qualities *(Cont.)*

Planning Sheet

Select a classmate for whom you will acknowledge some positive qualities. Design and present an award.

Name of this person: _____

What are the positive qualities of this person that you want to acknowledge?

_____

_____

_____

_____

Blue Ribbon Award
Describing Positive Qualities *(Cont.)*

Sample

Blue Ribbon goes to

Rosa Garcia

for

being helpful and friendly.

What a Character!
Character Development Activity

This Project

In this project your students will create or describe a real life character. They may describe a character from a story they read or develop a character in a story they are going to write.

Note: If you use the template (*Character*) provided on the CD-ROM, direct your students to follow the prompts to fill in the requested information.

Computer Skills

- word processing
- changing the page setup
- using ovals and text boxes
- entering data
- using the **Fill** and **Line** tools

Before Beginning

- Students should learn about the many facets that comprise a character such as physical description, personality, strengths and weaknesses.
- They should complete the planning sheet on Page 13.

Quick Steps

- Open a new *Word* document.
- Change the page orientation to **Landscape**.
- Insert one oval at the center of the page.
- Insert five ovals around the first oval.
- Fill in the ovals with colors.
- Draw a line from the center oval to each of the outer ovals.
- Insert text boxes inside the ovals.
- Fill in the requested data in each text box.
- Save and print your work.

What a Character!
Character Development Activity *(Cont.)*

Step 1 Open a new *Word* document.

Step 2 From the **FILE** menu, select **Page Setup**. Choose the **Landscape** Orientation in the pop-up window. Click **OK**.

Step 3 From the **VIEW** menu, choose **Toolbars**, then make sure **Formatting** is checked.

Step 4 From the **INSERT** menu, choose **Picture**, then **AutoShapes** from the drop-down menu. (If using a version of *Microsoft Word* prior to *97/98*, simply choose the **Circle** from the **Drawing Toolbar**.)

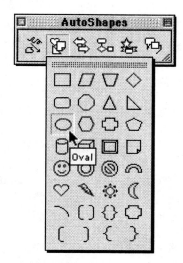

Step 5 In the **AutoShapes** pop-up window, select the second option, **Basic Shapes**, then the **Oval**. Click on the page and then drag one of the handles to insert the oval. Move the oval by clicking inside and dragging it to the center of the page.

Step 6 Insert five more ovals. Size and move them around the center oval.

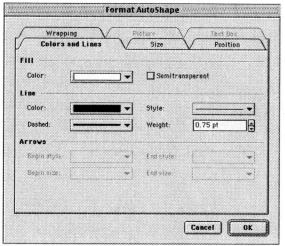

Step 7 To fill the ovals with colors, click on one oval at a time. The **Format AutoShapes** window will pop up. Click the **Colors and Lines** tab. Choose a new fill color.

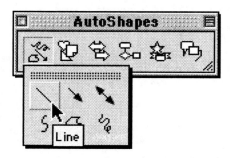

Step 8 In the **AutoShapes** window, click on the first option, **Lines**. Select the straight **Line**. Draw a line from the center oval to each of the five outer ovals.

What a Character!
Character Development Activity *(Cont.)*

Step 9   From the **INSERT** menu, select **Text Box**. Drag to draw a text box inside the center oval.

Step 10   Type the name of your character in the text box.

Step 11   Insert a text box inside one of the outer ovals.
Type "**Physical description:**"
Then enter the requested data.
Highlight "**Physical description:**" Click on the **Bold** button.   **B** *I* <u>U</u>

Step 12   Insert a text box inside the next outer oval. Type "**Personality:**" Then enter the requested data. Highlight "**Personality:**" Click on the **Bold** button.

Step 13   Insert a text box inside the next outer oval. Type "**Character's strengths and/or weaknesses:**" Then enter the requested data. Highlight "**Character's strengths and/or weaknesses:**" Click on the **Bold** button.

Step 14   Insert a text box inside the next outer oval. Type "**What difficulties does this character overcome?**" Then enter the requested data. Highlight "**What difficulties does this character overcome?**" Click on the **Bold** button.

Step 15   Insert a text box inside the next outer oval. Type "**Character's actions:**" Then enter the requested data. Highlight "**Character's actions:**" Click on the **Bold** button.

Step 16   Save and print your work.

What a Character!
Character Development Activity *(cont.)*

Planning Sheet

Select a character from history, a story, or someone you know in your real life.

What is your character's name?

_____

Describe your character's physical characteristics. For example, is he or she tall or short, stocky or slim? What color is the character's hair, eyes, skin?

_____

_____

Describe your character's personality. Is he or she generous or stingy, hard-working or lazy, friendly or mean?

_____

_____

What strengths and/or weaknesses does your character have?

_____

_____

What problems does your character have to overcome?

_____

What are some of your character's actions? What does he or she do in history, the story, or real life?

_____

_____

What a Character!
Character Development Activity *(Cont.)*

Template

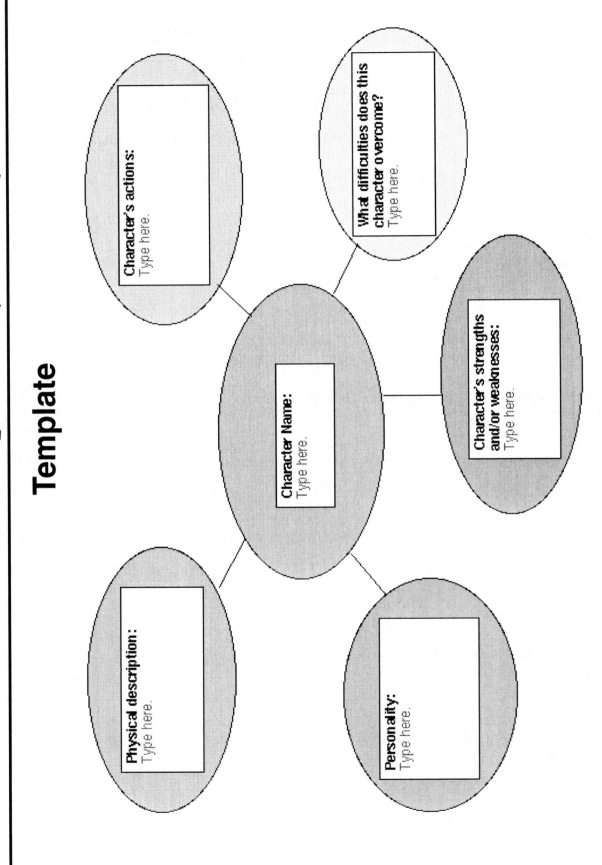

Character's actions:
Type here.

What difficulties does this character overcome?
Type here.

Character Name:
Type here.

Character's strengths and/or weaknesses:
Type here.

Physical description:
Type here.

Personality:
Type here.

What a Character!
Character Development Activity *(Cont.)*

Sample

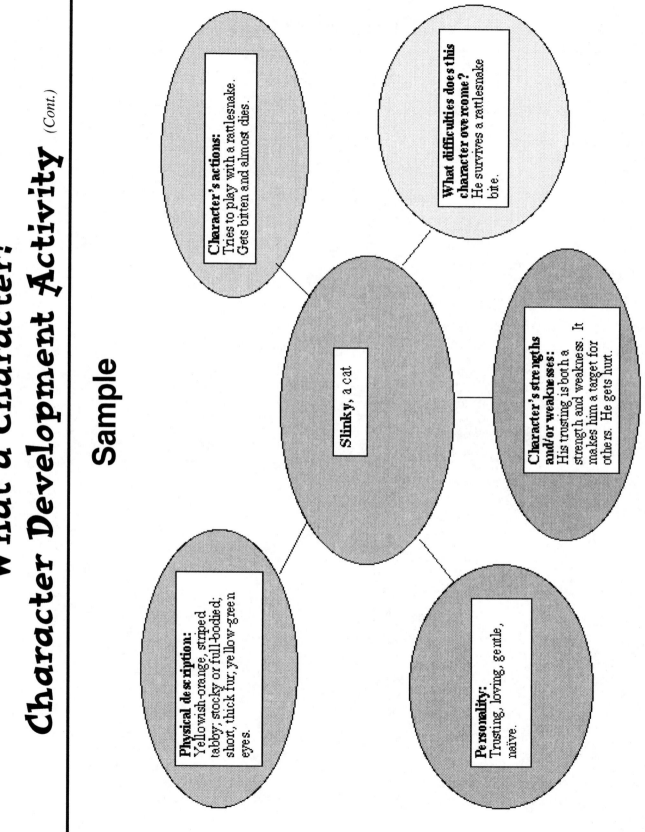

Character's actions:
Tries to play with a rattlesnake.
Gets bitten and almost dies.

What difficulties does this character overcome?
He survives a rattlesnake bite.

Slinky, a cat

Character's strengths and/or weaknesses:
His trusting is both a strength and weakness. It makes him a target for others. He gets hurt.

Physical description:
Yellowish-orange, striped tabby; stocky or full-bodied; short, thick fur; yellow-green eyes.

Personality:
Trusting, loving, gentle, naïve.

You Don't Say!
Dialogue Activity

This Project

In this project your students will create and write a dialogue to accompany clip art characters.

Note: You must have *Word 98/Office 97* or higher to use the template (*Dialogue*) for this lesson. Direct your students to follow the prompts to fill in the requested information.

Computer Skills

- word processing
- importing graphics from clip art
- using **Callout** bubbles from **AutoShapes**

Before Beginning

Students should discuss the use of dialogues in creative writing. See the sample on Page 20.

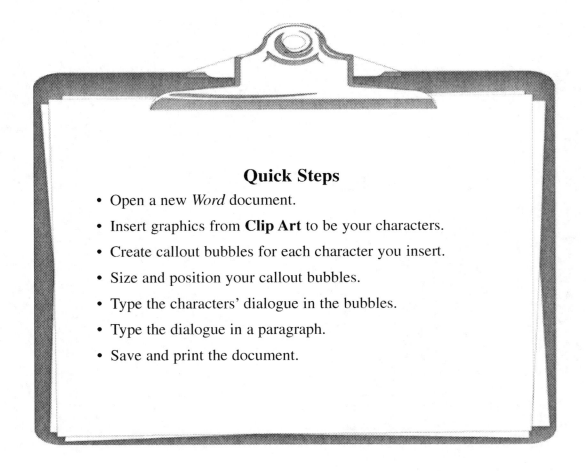

Quick Steps

- Open a new *Word* document.
- Insert graphics from **Clip Art** to be your characters.
- Create callout bubbles for each character you insert.
- Size and position your callout bubbles.
- Type the characters' dialogue in the bubbles.
- Type the dialogue in a paragraph.
- Save and print the document.

You Don't Say!
Dialogue Activity *(Cont.)*

Step 1   Open a new *Word* document.

Step 2   Click on the **Center** button.

Select a large font size (**22** to **26** points.)

Type the title of your dialogue.

Step 3   Reset the font size to **12** points.

Press the **Enter** or **Return** key.

Click on the **Align Left** button.

Press the **Enter** or **Return** key about 15 times.

Step 4   From the **INSERT** menu, choose Picture, then **Clip Art** from the drop-down menu.

Step 5   From the **Microsoft Clip Gallery**, select a category. Choose a graphic. Click on the Insert button.

Step 6   Size the graphic by dragging the handles around it. Move the graphic by clicking inside and dragging it.

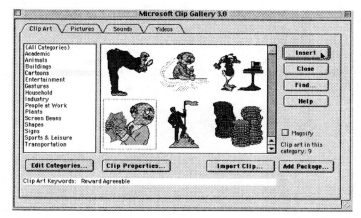

Step 7   From the **INSERT** menu, choose Picture, then **AutoShapes** from the drop-down menu.

Step 8   From **AutoShapes**, select the last option, **Callouts**. From the pop-up window, choose a callout bubble.

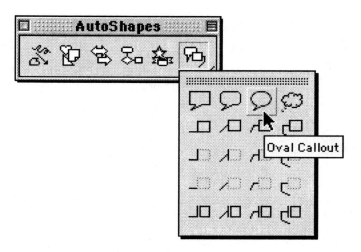

You Don't Say!
Dialogue Activity *(Cont.)*

Step 9 Click above your character to insert the callout bubble. Size the bubble by dragging the handles around it. Move the bubble by clicking inside and dragging it.

Step 10 If you click on the pointer arrow, a yellow box will appear. You can drag this yellow box to change the direction of the pointer arrow so that it points to your character.

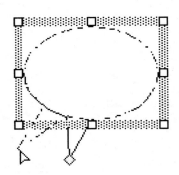

Step 11 Repeat **Steps 4** to **6** to insert more graphics.

Step 12 Repeat **Steps 7** to **10** to insert a callout bubble for each of your characters.

Step 13 Click inside each bubble. Type your dialogue.

Step 14 This is optional: click under the clip art, and type the dialogue in a paragraph.

Step 15 Save and print the document.

You Don't Say!
Dialogue Activity *(Cont.)*

Template

Title of Your Dialogue

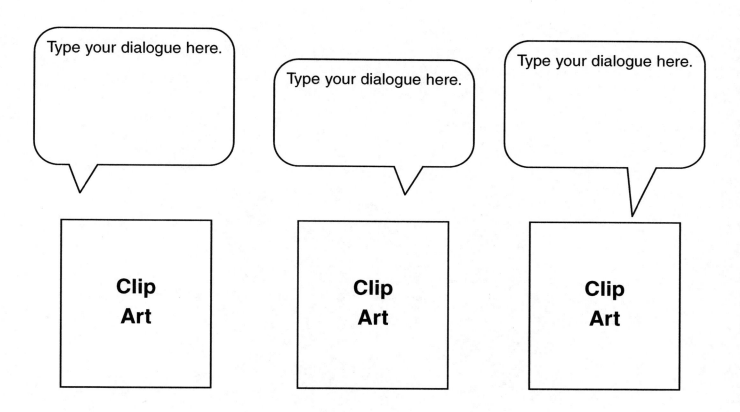

Type your dialogue in a paragraph here. (Optional)

You Don't Say!
Dialogue Activity *(Cont.)*

Sample

It's a Pleasure!

"It's a deal, Roberta. It's a pleasure working with you," exclaimed Jack.

Alex responded, "Count me in, too. How did you think of that?"

Roberta replied, "The pleasure is mine. Glad I could help out. Let me know if I can be of more help."

5-7-5
Haiku Word Art

This Project

In this project your students will write and present a haiku using **WordArt**. **WordArt** is a function within *Word* which helps you present your writing in an artistic and colorful design. This project is suitable for many types of short poems.

Computer Skills

- word processing
- using the **WordArt** tool

Before Beginning

Students should write a haiku using the planning sheet on Page 23.

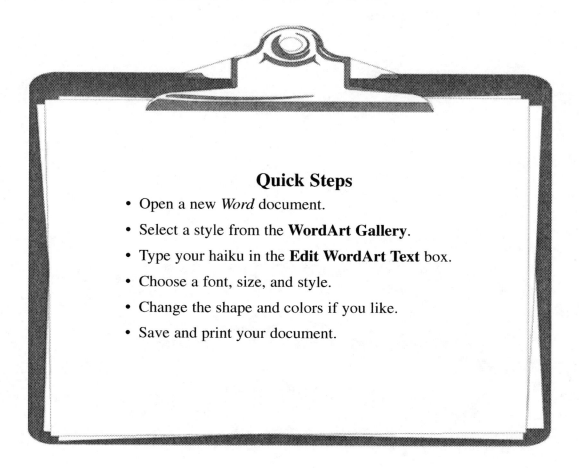

Quick Steps

- Open a new *Word* document.
- Select a style from the **WordArt Gallery**.
- Type your haiku in the **Edit WordArt Text** box.
- Choose a font, size, and style.
- Change the shape and colors if you like.
- Save and print your document.

5-7-5
Haiku Word Art *(Cont.)*

Step 1   Open a new *Word* document.

Step 2   From the **INSERT** menu, select **Picture**, then **WordArt** form the drop-down menu.

Step 3   Select a style in the **WordArt Gallery** that pops up. Click **OK**.

Step 4   Type your haiku in the **Edit WordArt Text** box.

Step 5   Select a font, size, and style. Click **OK**.

Step 6   If you like, you can change the size and the shape by dragging the handles around your haiku.

Step 7   To change the colors of your haiku, select the haiku. From the **FORMAT** menu, select **WordArt**. The Format **WordArt** window will pop up. Click the **Colors and Lines** tab. Choose a new fill color. Click **OK**.

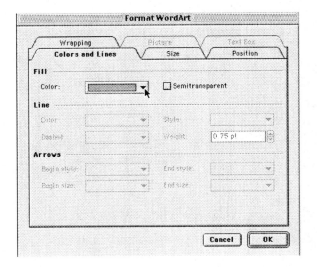

Step 8   Save and print your document.

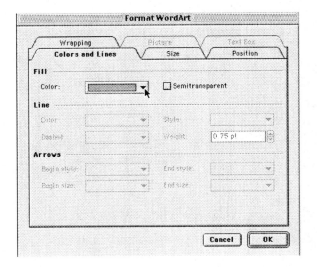

Japanese haiku
Captures a moment in time
Snapshot memory

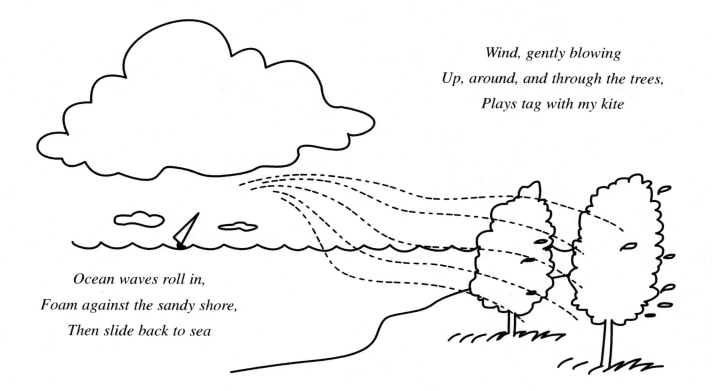

5-7-5
Haiku Word Art *(Cont.)*

Planning Sheet

A haiku is a poem of three lines. The first line has five syllables. The second has seven. The third has five.

First line _____

Second line_____

Third line_____

Wind, gently blowing
Up, around, and through the trees,
Plays tag with my kite

Ocean waves roll in,
Foam against the sandy shore,
Then slide back to sea

5-7-5
Haiku Word Art *(Cont.)*

Sample

It's in the Mail!
Letterhead and Letter Writing

This Project

In this project your students will design a letterhead and write a business letter. This could be an actual letter they will send to request materials for a report. It could also be an imaginary letter that a character in a story might write, or a letter to a character in a story.

Computer Skills

- word processing
- selecting fonts, styles, and sizes
- coloring and aligning text

Before Beginning

Students should view letterhead samples and learn the form elements of business letters. It is now common in business letters to have all the paragraphs, greeting, and signature aligned at the left margin (flush left.) Students can use the planning sheet on Page 27.

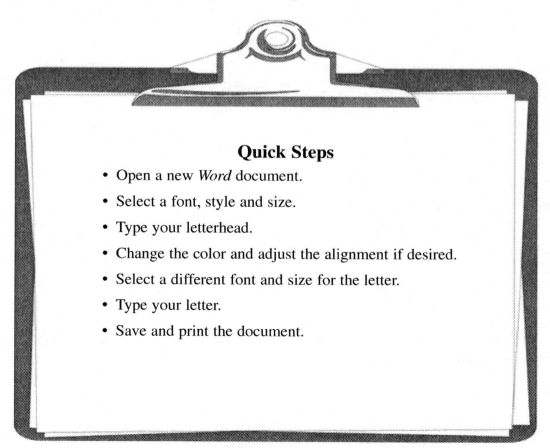

Quick Steps

- Open a new *Word* document.
- Select a font, style and size.
- Type your letterhead.
- Change the color and adjust the alignment if desired.
- Select a different font and size for the letter.
- Type your letter.
- Save and print the document.

It's in the Mail!
Letterhead and Letter Writing *(Cont.)*

Step 1 Open a new *Word* document.

Step 2 On the **Formatting Toolbar**, select a font, font size and style.

Step 3 Type your name, address, phone number and other information as they should appear on the letterhead.

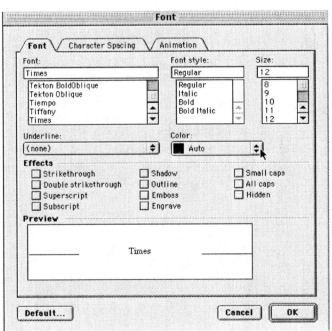

Step 4 To change the color of the letterhead, highlight it by dragging the cursor across it. From the **FORMAT** menu, select **Font**. In the **Font** dialogue box that pops up, click the **Font** tab. Choose a new color. Click **OK**.

Step 5 Align the letterhead by clicking on the **Align Left**, **Align Right** or **Center** button. Click at the end of the letterhead. Press the **Enter** or **Return** key twice.

Step 6 Select a different font and font size. Click on the **Align Left** button.

Step 7 Type your letter.

Step 8 Save and print your document.

It's in the Mail!
Letterhead and Letter Writing *(Cont.)*

Planning Sheet

For the letterhead:

Write the name, address, phone number and any other information to appear on the letterhead. Sketch your word placement and sizing ideas.

For the letter:

Write the name and the address of the person to whom you are writing the letter.

_____

_____

_____

_____

Write a draft of your letter.

_____

_____

_____

_____

_____

_____

_____

_____

_____

It's in the Mail!
Letterhead and Letter Writing *(Cont.)*

Sample

Linda Pino

12 My Street, San Francisco, CA 94127 (415) 555-1212

February 1, 2000

President William Clinton
1600 Pennsylvania Avenue
Washington, D.C. 20500

Dear President Clinton:

I am in the fifth grade and am writing a report on the White House. Could you please send me pictures, maps, and articles about the White House? I am most interested in Socks' room.

Thank you very much.

Sincerely,

It's Negotiable
A Conflict Resolution Activity

This Project

In this project your students will write about resolving a conflict. They will identify the groups in the conflict and list each group's interests. They will evaluate possible resolutions. The conflict might be real or imaginary, from history or an actual conflict that they or their classmates are confronting.

Note: If you use the template (*Negotiable*) provided on the CD-ROM, direct your students to follow the prompts in the document.

Computer Skills

- word processing
- creating a table
- using the bullets and numbering tool
- using different font styles

Before Beginning

- Identify a conflict, the groups in the conflict, and each group's interests. Discuss conflict resolution and how to try to meet the needs of each group. Students can use the planning sheet on Page 31.

Quick Steps

- Open a new *Word* document.
- Type "**A Conflict Resolution**" as your title.
- Write a paragraph about a problem or conflict.
- Create a table of two columns and one row.
- Type the names of the two groups in the conflict and list their interests.
- Write about what might happen if the conflict is not resolved.
- List several resolutions to the conflict.
- Write about the best resolution and explain why.
- Save and print your document.

It's Negotiable
A Conflict Resolution Activity *(Cont.)*

Step 1   Open a new *Word* document.

Step 2   Click on the **Center** button on the **Formatting Toolbar**.
Set the font size to **24** or **36** points.
Type "**A Conflict Resolution**".

| Title | ▼ | Times | ▼ | 24 | ▼ |

Step 3   Reset the font size to **12** points. Press the **Enter** or
Return key twice. Click on the **Align Left** button.
Type a paragraph about a conflict or problem. Press
the **Enter** or **Return** key twice.

Step 4   From the **TABLE** menu, select **Insert Table**. In the pop-up window, type in **2** columns, **1**
row, and select **Auto** column width. Click **OK**.

Step 5   Click at the top of the first column. Type "**Group 1:**" and the name of the first group in the
conflict. Press the **Enter** or **Return** key.

Step 6   Type the first group's interests.

Step 7   Click at the top of the second column. Type "**Group 2:**" and the name of the second group in
the conflict. Press the **Enter** or **Return** key.

Step 8   Type the second group's interests.

Step 9   Click under the table. Press the **Enter** or **Return** key.
Click on the **Bold** button on the **Formatting Toolbar**.
Type "**What will happen if the conflict isn't resolved?**"
Press the **Enter** or **Return** key. Click on the **Bold** button again to deactivate the function.
Type your answer to the question. Press the **Enter** or **Return** key twice.

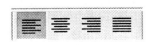

Step 10   Click on the **Bold** button. Type "**Possible conflict resolutions:**" Press the **Enter** or **Return**
key. Click on the **Bold** button again to deactivate the function. List several possible ways to
resolve the conflict. Press the **Enter** or **Return** key at the end of each resolution.

Step 11   Highlight the list by dragging the cursor across it. From the **FORMAT** menu, select **Bullets
and Numbering**. Click on the **Numbered** tab in the pop-up window. Choose a style and
click on the **OK** button. Press the **Enter** or **Return** key twice.

Step 12   Click on the **Bold** button. Type "**What is the best resolution and why?**" Press the **Enter** or
Return key. Click on the **Bold** button again to deactivate the function. Type what you think
is the best resolution and explain why.

Step 13   Save and print the document.

It's Negotiable
A Conflict Resolution Activity *(Cont.)*

Planning Sheet/Template

Problem:

Group 1: Interests:	Group 2: Interests:

What will happen if the conflict isn't resolved?

_____

_____

Possible conflict resolutions:

_____

_____

_____

What is the best resolution and why?

_____

_____

_____

It's Negotiable
A Conflict Resolution Activity *(Cont.)*

Sample

Problem: There is only one computer at the back of the room. Though the students in Mr. Garcia's fourth-grade class usually have specific assignments, this week they are allowed "free time", or an activity of their choice, during their center rotation time. The only requirement is that they work in pairs.

Hector and Seth can't agree on what to do. They only have 20 minutes.

Group 1: Hector **Interests:** play football math game	**Group 2:** Seth **Interests:** draw with *Kid Pix*

What will happen if the conflict isn't resolved?

Neither Hector nor Seth will do what they want to do at the computer. Maybe neither will use the computer at all, or one will have to do what the other wants.

Possible conflict resolutions:

- Hector agrees to do what Seth wants.
- Seth agrees to do what Hector wants.
- Hector plays his game for 10 minutes. Seth draws for 10 minutes.
- Seth and Hector decide to stay at their seats and not use the computer.
- Seth does what he wants this week. Hector gets his way next time there is free time, or vice versa.

What is the best resolution and why?

Number 3 is the best resolution because then both Hector and Seth do what they each want to do for their allotted time. Since they don't get free time each week, resolution 5 might not work. In resolutions 1 and 2, someone is not getting to do what they want. Therefore, resolution 3 seems the fairest. It meets both groups' interests.

Putting Things in Order
A Time Line Project

This Project

In this project your students will create a time line to represent a person's life or a historical era. They will highlight eight events in this person's life or this historical era after researching the biography or topic.

Computer Skills

- word processing
- entering data
- drawing straight lines
- inserting, sizing and moving text boxes

Before Beginning

- Have students review the elements of a time line. They should research a selected topic or person in history, then choose a time period for the time line and divide it into 10 equal segments.

- Students select eight of the most important events that took place during this time period and complete the planning sheet on Page 35.

Quick Steps

- Open the template called *"Timeline"*.
- Fill in the title of your time line.
- Type the first year of each time segment on the time line.
- Type the year and a brief description of each event in the text boxes.
- Size and move the text boxes if necessary.
- Draw a straight line from each text box to the point of reference on the time line.
- Save and print your work.

Putting Things in Order
A Time Line Project *(Cont.)*

Step 1 Open *Microsoft Word*. From the **FILE** menu, select **Open**.

Step 2 Find the file called *"Timeline"*. Double-click on it to open the file.

Step 3 Click after the word **"of"** in the title. Type your topic of the time line.

Step 4 Click in the first cell in the table. Type the first year of the first time period.

1910									

Step 5 Fill in the rest of the cells in the table with the beginning year of each time period.

Step 6 Click inside the first text box. Type the specific year when the first important event took place. Press the **Enter** or **Return** key. Type a brief description of the event.

Step 7 Size and move the text box if necessary. Size the box by clicking on it and dragging the handles around it. Move the box by clicking inside and dragging it.

Step 8 From the **INSERT** menu, select **Picture**, and then **AutoShapes** from the drop-down menu.

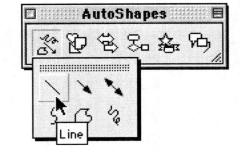

Step 9 In the **AutoShapes** window, select the **Line** tool. Draw a straight line from the text box to the relevant point on the time line.

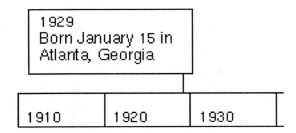

Step 10 In the rest of the text boxes, type the year and a brief description of each of the other seven events. Size and move the text boxes if necessary.

Step 11 Repeat **Step 9** to finish drawing lines from the text boxes to the time line.

Step 12 Save and print your work.

Putting Things in Order
A Time Line Project *(Cont.)*

Planning Sheet

Your topic:

The time period divided into 10 equal segments for the time line:

The eight most important events:

Year_____ Event_____

_____

Year_____ Event_____

_____

Year_____ Event_____

_____

Year_____ Event_____

_____

Year_____ Event_____

_____

Year_____ Event_____

_____

Year_____ Event_____

_____

Year_____ Event_____

_____

© *Teacher Created Resources, Inc.*   35   #2728 Simple Projects for Microsoft® Word

Putting Things in Order A Time Line Project *(Cont.)*

Template

Time Line of Your Topic

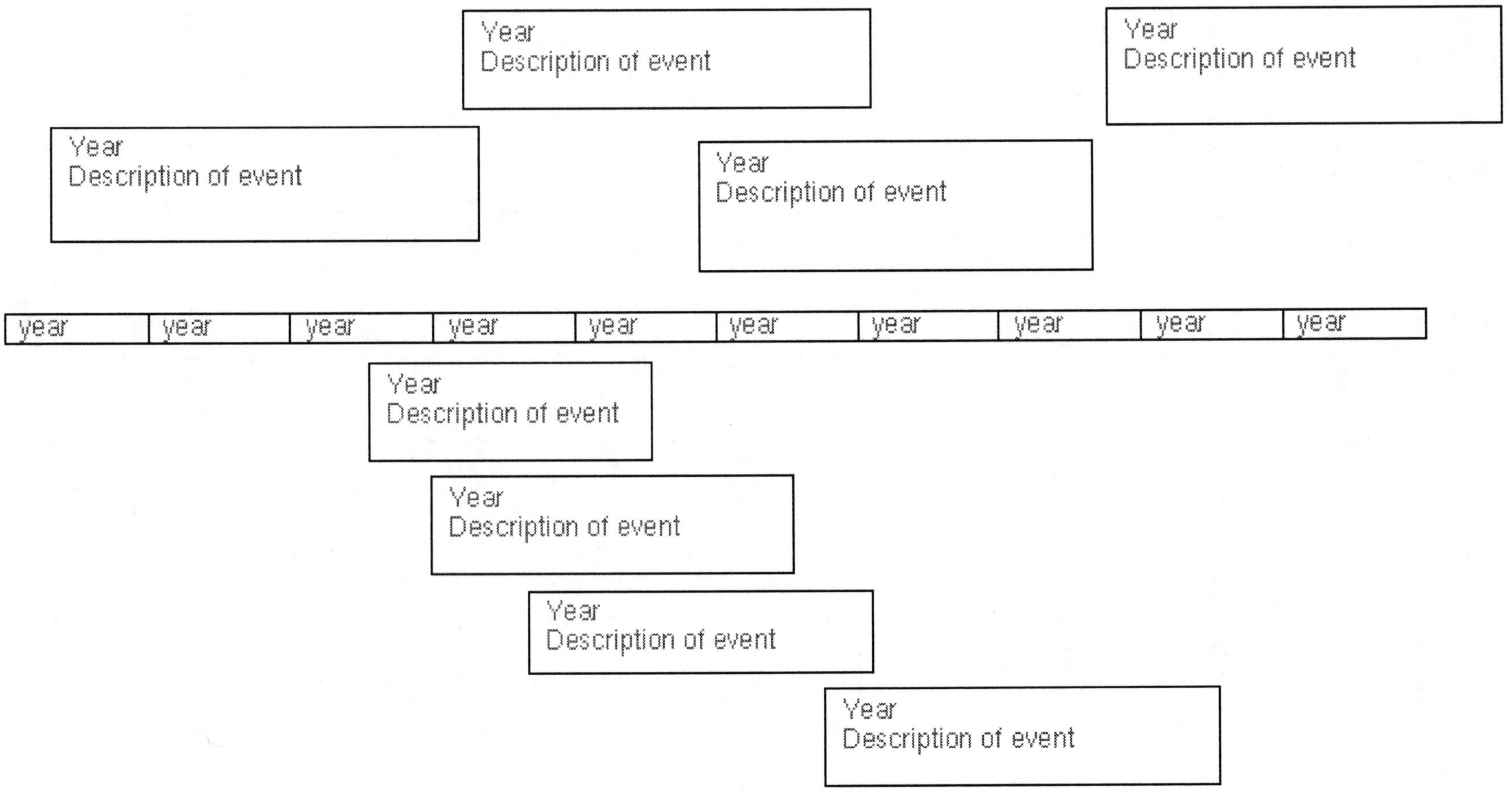

Year
Description of event

Year
Description of event

Year
Description of event

Year
Description of event

Year
Description of event

Year
Description of event

Year
Description of event

Year
Description of event

year
year
year
year
year
year
year
year

Putting Things in Order A Time Line Project (Cont.)

Sample

Time Line of Dr. Martin Luther King, Jr.

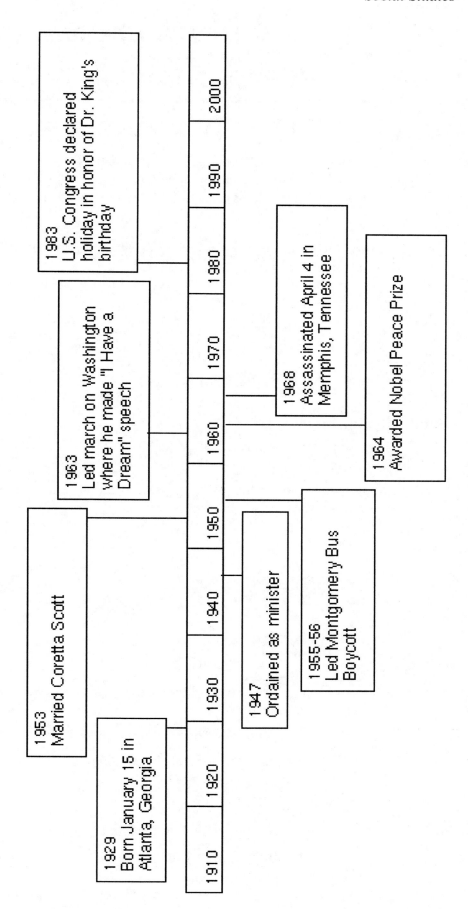

Culture Comparison
Native American Cultures

This Project

In this project your students will compare and contrast various Native American tribes. They will briefly describe the type of shelter, diet, and clothing used by each tribe and include other facts of interest. They will also write a brief paragraph about why shelters differ between the Native American tribes.

Note: If you use the template (*Culture*) provided on the CD-ROM, direct your students to fill in the information in the table and type a paragraph under the table.

Computer Skills

- word processing
- creating a table
- adjusting column width and row height
- entering data

Before Beginning

- Students should research two Native American tribes and collect information on their shelters, diet, clothing, and where they live.

- Distribute two copies of the planning sheet to each student (Page 40.) Students will complete one sheet for each of the tribes they studied.

Quick Steps

- Open a new *Word* document.
- Type "**Native American Culture Comparison Table**" as your title.
- Insert a table of **6** columns and **3** rows.
- Type the categories in the top row.
- Enter the requested information in the middle and bottom rows.
- Type a paragraph to explain about the two tribes' shelters.
- Save and print your document.

Culture Comparison
Native American Cultures *(Cont.)*

Step 1   Open a new *Word* document.

Step 2   Click on the **Center** button on the
Formatting Toolbar.

Set the font size to **22** points.

Click on the **Bold** button.

Type "**Native American Culture Comparison Table**" as your title.

Step 3   From the **TABLE** menu, select Insert Table. In the dialog box that pops up, type in **6**
columns, **3** rows, and choose Auto column width. Click **OK**.

Step 4   Highlight the top row by clicking to the left of the first cell.
Change the font size to **16**
points. Click in the first cell in
the top row. Type "**Tribe**".
Type "**Region**" in the second
cell, "**Shelter**" in the third,
"**Diet**" in the fourth, "**Clothing**" in the fifth, and "**Other**" in the last cell.

Step 5   Click to the left of the first row. Hold down the **Shift** key. Click to the left of the second row.
Both the middle and the bottom rows are selected now. Change the font size to **12** points.
Click on the **Bold** button to deactivate the function.

Step 6   Click in the first cell in the middle row. Type the name of the first tribe. Enter the first tribe's
region information in the second cell, shelter information in the third, diet information in the
fourth, clothing information in the fifth, and other facts of interest in the last cell.

Step 7   In the bottom row, type the name of the second tribe in the first cell. Enter the second tribe's
region information in the second cell, shelter information in the third, diet information in the
fourth, clothing information in the fifth, and other facts of interest in the last cell.

Step 8   Click under the table. Press the **Enter** or **Return** key.
Choose a **12**-point font size. Click on the **Align Left**
button.
Click on the **Bold** button to deactivate the function. Type a paragraph to explain why the
shelters differed between the two tribes.

Step 9   Save and print the document.

Culture Comparison
Native American Cultures *(Cont.)*

Planning Sheet

Use this chart to identify the types of shelter, diet, clothing, and other interesting facts for each Native American culture tribe or tribe that you select or are assigned.

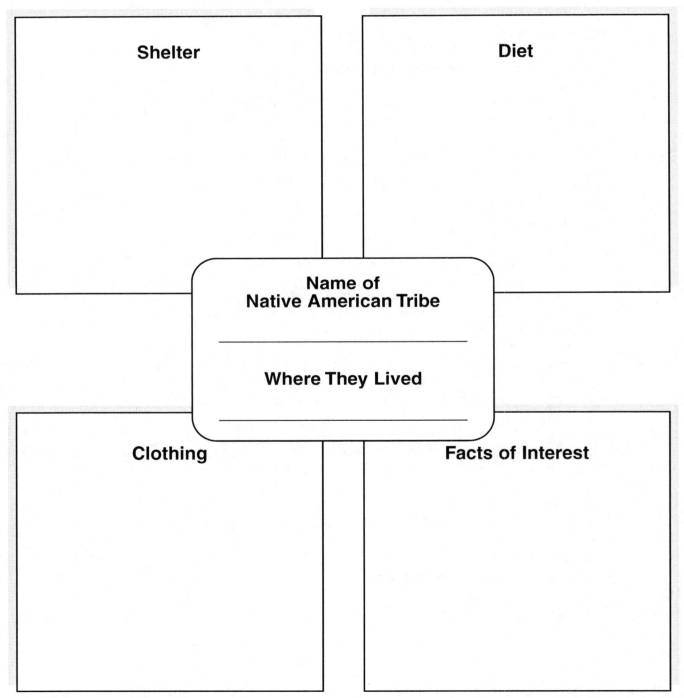

Shelter

Diet

**Name of
Native American Tribe**

_____

Where They Lived

_____

Clothing

Facts of Interest

Culture Comparison
Native American Cultures *(Cont.)*

Sample

Native American Culture Comparison Table

Tribe	Region	Shelter	Diet	Clothing	Other
Wampanoag	Eastern Woodlands	longhouses and wigwams	deer, rabbit, squirrel berries	hides of small animals	la crosse, wampum, weaving
Cheyenne	Plains	teepees	buffalo	buffalo hides	war bonnets

The information for this chart was obtained from the Web site at
http://www.germantown.k12.il.us/html/culture.html

Each Native American culture had their own type of home because of the natural resources that were found in the different environments. Wood was available in the forests, so Native Americans in the Northwest and Eastern Woodlands were able to build wooden lodges and wigwams. Teepees were built in the plains because of the large buffalo skins that were available.

Geography Term Shuffle Vocabulary Activity

This Project

In this project your students will match geography terms with their definitions and give an example of each. This project is easily adaptable for use with vocabulary in any of the content areas.

Computer Skills

- word processing
- moving text boxes

Before Beginning

- Students should be familiar with geography terms such as archipelago, canal, canyon, cape, continent, gulf, isthmus, peninsula, and valley.

- Students should use an atlas to identify an example of each of the terms.

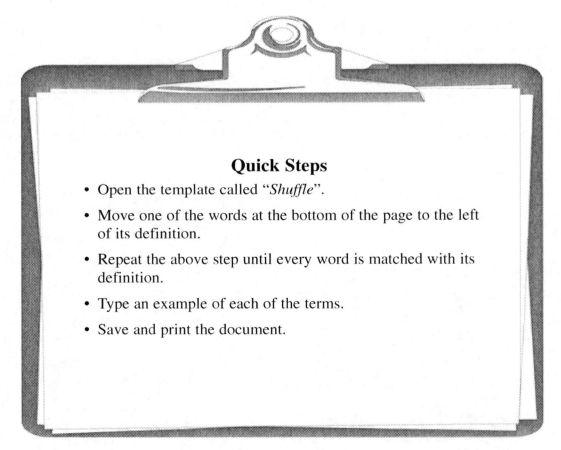

Quick Steps

- Open the template called "*Shuffle*".

- Move one of the words at the bottom of the page to the left of its definition.

- Repeat the above step until every word is matched with its definition.

- Type an example of each of the terms.

- Save and print the document.

Geography Term Shuffle
Vocabulary Activity *(Cont.)*

Step 1 Open *Microsoft Word*. From the **FILE** menu, select **Open**.

Step 2 Find the file called *"Shuffle"*. Double-click on it to open the file.

Step 3 Click on one of the terms at the bottom of the page to highlight it. Drag the box to the left of its definition.

Step 4 Repeat **Step 3** to match every term with its definition.

Step 5 Click in the second page. Follow the prompts to type an example for each of the terms.

Step 6 Save and print your document.

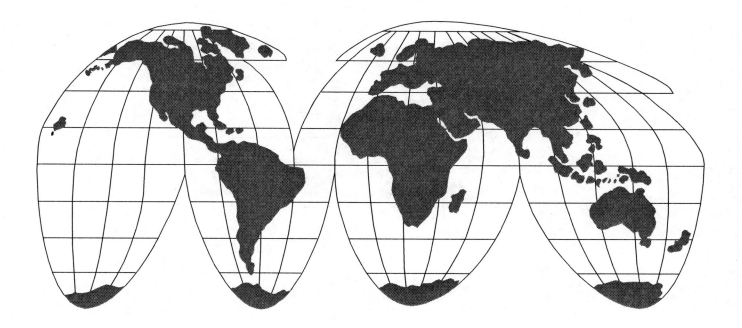

Geography Term Shuffle
Vocabulary Activity *(Cont.)*

Template 1

Drag each of the geography terms to the left of its definition.

One of the largest land masses on earth

Low area that is surrounded by hills or mountains

Man-made waterway

A deep valley with steep sides

A point of land jutting into the water

Area of land that is surrounded by water on all sides, except one

Part of the ocean that is enclosed by land

Narrow strip of land that connects two larger land areas

Large group of islands in the sea

Isthmus   **Archipelago**   **Gulf**   **Peninsula**   **Valley**

Canyon   **Continent**   **Cape**   **Canal**

Geography Term Shuffle
Vocabulary Activity *(Cont.)*

Template 2

Provide a real-world example of each term listed below. For example: Continent: North America.

Continent:

Valley:

Canal:

Canyon:

Cape:

Peninsula:

Gulf:

Isthmus:

Archipelago:

Geography Term Shuffle
Vocabulary Activity *(Cont.)*

Answer Sheet 1

Continent	One of the largest land masses on earth
Valley	Low area that is surrounded by hills or mountains
Canal	Man-made waterway
Canyon	A deep valley with steep sides
Cape	A point of land jutting into the water
Peninsula	Area of land that is surrounded by water on all sides, except one
Gulf	Part of the ocean that is enclosed by land
Isthmus	Narrow strip of land that connects two larger land areas
Archipelago	Large group of islands in the sea

Geography Term Shuffle
Vocabulary Activity *(Cont.)*

Sample Answers for Template 2

Continent: South America

Valley: Salinas Valley, California

Canal: Panama Canal

Canyon: Grand Canyon

Cape: Cape Cod, Massachussetts

Peninsula: Monterey Peninsula, CA

Gulf: Gulf of Mexico

Isthmus: Isthmus of Suez

Archipelago: Malaysia

Floor Plan
Classroom Map

This Project

In this project your students will use shapes to create a classroom map.

Note: You must have Word *98/Office 97* or higher to do this lesson.

Computer Skills

- word processing
- inserting shapes from **AutoShapes**
- copying and pasting shapes
- filling shapes with color

Before Beginning

- Distribute a copy of the planning sheet to each student. Students will list items in the classroom and note the location of each item.
- Students should sketch a map of their classroom using different shapes to represent the items in their classroom. Use the planning sheet on Page 50.

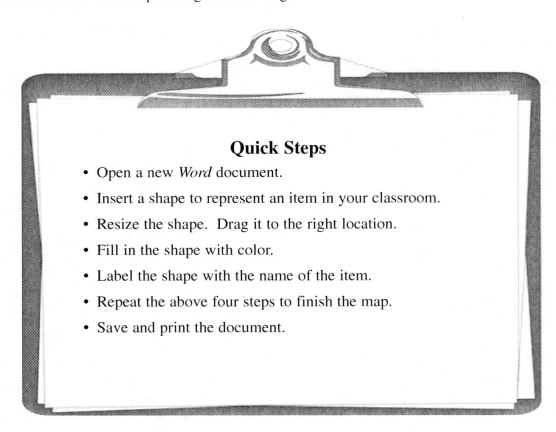

Quick Steps

- Open a new *Word* document.
- Insert a shape to represent an item in your classroom.
- Resize the shape. Drag it to the right location.
- Fill in the shape with color.
- Label the shape with the name of the item.
- Repeat the above four steps to finish the map.
- Save and print the document.

Floor Plan
Classroom Map *(Cont.)*

Step 1   Open a new *Word* document.

Step 2   From the **INSERT** menu, choose **Picture**, then **AutoShapes** in the drop-down menu.

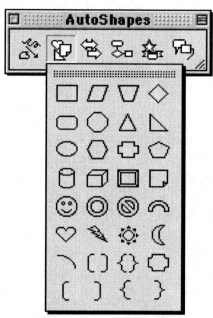

Step 3   From **AutoShapes**, select the second option, **Basic Shapes**. From the pop-up window, choose a shape.

Step 4   Click to insert the shape. Size the shape by dragging the handles around it. Move the shape by clicking inside and dragging it to the desired location on the page.

Step 5   Double-click on the shape. In the **Format AutoShape** window, click the **Colors and Lines** tab. Choose a new fill color. Click **OK**.

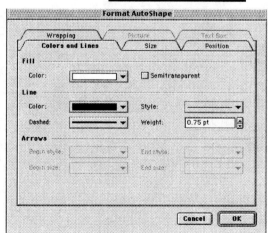

Step 6   From the **INSERT** menu, select **Text Box**. Click and drag the cursor inside the shape. Type in the text box to label the item.

Step 7   Repeat **Steps 3–6** to finish representing the other items in the classroom on your map. (To use a shape more than once, click inside to highlight that shape. From the **EDIT** menu, select **Copy**. Click outside of the shape. Select **Paste** from the **EDIT** menu. While the shape is still highlighted, move it to the desired location. Size it if necessary.)

Step 8   Save and print the document.

Floor Plan
Classroom Map *(Cont.)*

Planning Sheet

List and count the items to be included in your classroom map. Draw a simple shape to represent each of these items.

Item	**Number**	**Shape to represent the item**
Desks (student)	_____	
Desks (teacher)	_____	
Shelves (large)	_____	
Shelves (small)	_____	
Tables	_____	
Rugs	_____	
Other	_____	
Other	_____	
Other	_____	

On the other side of this sheet, sketch where you will place these items. Illustrate where the doors and the windows are. Show where other items or places of importance are, such as the fire extinguisher, the reading center, etc.

Floor Plan
Classroom Map *(Cont.)*

Sample

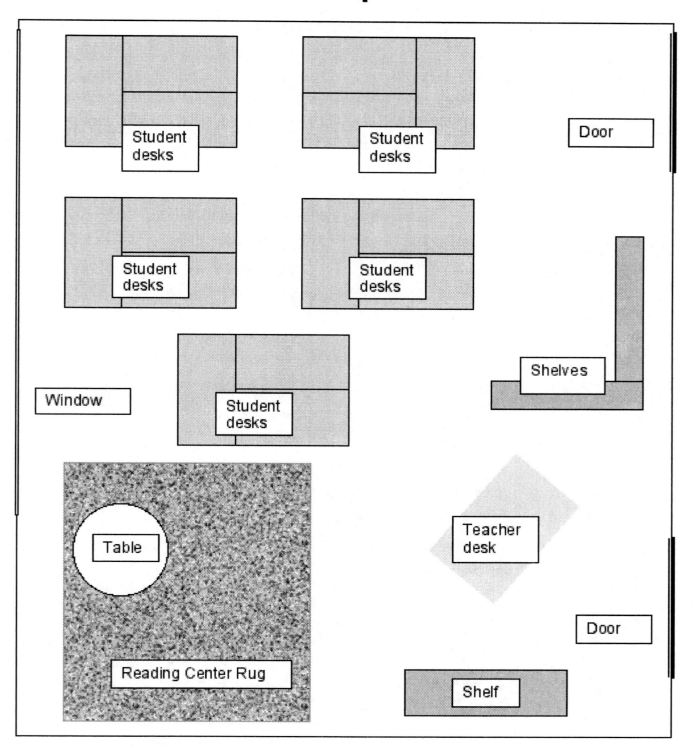

Campaign Promises
An Election Flyer

This Project

In this project your students will design an election flyer. They can cover an actual election in the United States or world history, current local, state or national elections, or student body elections.

Note: If you use the template (*Campaign*) provided on the CD-ROM, direct your students to follow the prompts to fill in the information.

Computer Skills

- word processing
- inserting text boxes
- using the bullet
- using the **Fill Color** tool

Before Beginning

- Students should learn about the important issues in an election and what makes an effective flyer or leaflet.
- They complete the planning sheet on Page 54.

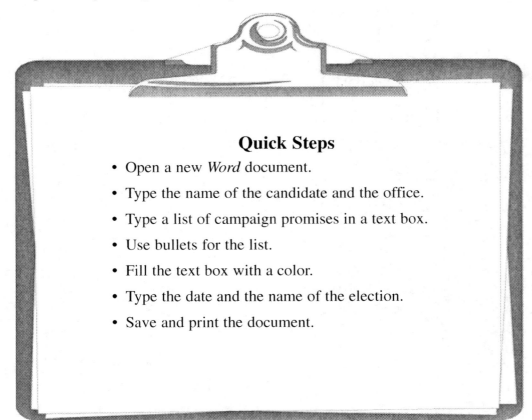

Quick Steps

- Open a new *Word* document.
- Type the name of the candidate and the office.
- Type a list of campaign promises in a text box.
- Use bullets for the list.
- Fill the text box with a color.
- Type the date and the name of the election.
- Save and print the document.

Campaign Promises
An Election Flyer *(Cont.)*

Step 1   Open a new *Word* document.

Step 2   Click on the **Center** button on the **Formatting Toolbar**.
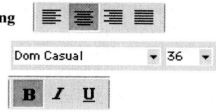
Choose a large font size (**36** or **48** points.)
Click on the **Bold** button.
Type the name of the candidate.
Press the **Enter** or **Return** key.

Step 3   Choose a slightly smaller font size. Click on the **Bold** button to deactivate the function. Type the name of the office for which you are voting. Press the **Enter** or **Return** key about 10 times.

Step 4   From the **INSERT** menu, select **Text Box**. Click under the name of the office to insert the text box. Resize the text box by dragging the handles around it. Move it by clicking inside and dragging it. The text box should bet in the center under the name of the office.

Step 5   Type a list of reasons for voting for the candidate. Press the **Enter** or **Return** key at the end of each reason.

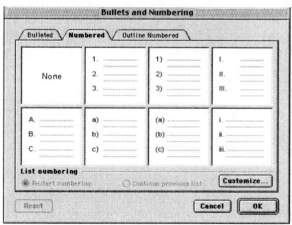

Step 6   Highlight the list by dragging the cursor across it. Change the font size to **22** or **24** points. From the **FORMAT** menu, select **Bullets and Numbering**. In the pop-up window, click on the **Bulleted** or **Numbered** tab to choose a style. Click **OK**.

Step 7   Double-click on one of the sides of the text box. In the **Format Text Box** window, click the **Colors and Lines** tab. Choose a new fill color. Click **OK**.

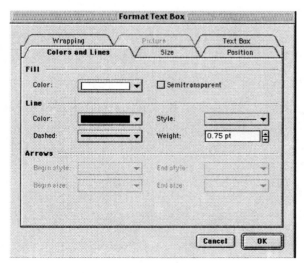

Step 8   Click under the text box. Type the date and name of the election.

Step 9   Add graphics or borders if desired.

Step 10   Save and print the document.

Campaign Promises
An Election Flyer *(Cont.)*

Planning Sheet

Name of the candidate

_____

Office for which the candidate is running

_____

Date and name of the election

_____

List the reasons why you would vote for this candidate.

1. _____

2. _____

3. _____

4. _____

Campaign Promises
An Election Flyer *(Cont.)*

Sample

Melissa Fair
for Student Body President

A Vote for Melissa is a vote for:

- Lunch-time club activities
- After-school dances
- Inter-class soccer games
- Reward movie assemblies

January 18
Student Body Elections

Koalas, Kangaroos, and More Animal Report

This Project

In this project your students will create a brief animal report. They will write about an animal and present data in a table. Students can also import a photo or drawing of the animal.

Note: If you use the template (*Animal*) provided on the CD-ROM, direct the students to follow the prompts to fill in the requested information.

Computer Skills

- word processing
- creating a table
- adjusting column width
- entering data
- importing graphic (optional)

Before Beginning

- Students research a selected animal and its characteristics, such as physical description, habitat, food, life span, reproduction and order.
- Students complete the planning sheet on Page 58.

Quick Steps

- Open a new *Word* document.
- Type the title of your report.
- Type a brief description of the animal.
- Create a table of six columns and two rows.
- Type your categories in the top row.
- Fill in the requested data in the bottom row.
- Import a photograph or drawing (optional).
- Save and print your document.

Koalas, Kangaroos, and More Animal Report *(Cont.)*

Step 1 Open a new *Word* document.

Step 2 Click on the **Center** button on the **Formatting Toolbar**.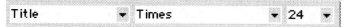

Set the font size to **24** points.

Type the title of your report.

Step 3 Reset the font size to **12** points. Press the **Enter** or **Return** key twice.

Click on the **Align Left** button.

Type one or more paragraphs about your animal.

Press the **Enter** or **Return** key twice.

Step 4 From the **TABLE** menu, select **Insert Table**. In the pop-up window, type in **5** for the number of columns, **2** for the number of rows, and choose **Auto** for the column width. Click **OK**.

Step 5 Click to the left of the top row to highlight it. Click on the **Bold** and then the **Center** button. Type one category in each of the cells in the top row.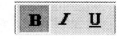

Step 6 Click on the **Bold** button again to deactivate the function. Type the requested information in each of the cells in the bottom row.

Step 7 If you need to adjust a column's width, move the cursor over the vertical line of the column. When the cursor changes to a resizing arrow, click and drag the line to the left or right.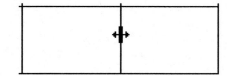

Step 8 To import a photo or drawing, click where you want to place the graphic. From the **INSERT** menu, select **Picture**, then **From File**, find the picture file you want to insert. Click on the filename. Then click on the **Insert** button.

Step 9 Size this graphic if necessary by dragging the handles around it.

Step 10 Save and print your document.

Koalas, Kangaroos, and More
Animal Report *(Cont.)*

Planning Sheet

Name of the animal: _____

Physical Description: _____

Habitat and Range: _____

Food: _____

Life Span: _____

Reproduction: _____

Write one or more paragraphs about the animal here.

_____

_____

_____

_____

_____

_____

_____

_____

_____

_____

_____

Koalas, Kangaroos, and More Animal Report *(Cont.)*

Sample

Sloths

Sloths are the slowest mammal in the world. They are usually about 2 feet (61 cm) long and live in trees. Sloths eat, sleep, and mate high above the ground.

Sloths sleep about 15 hours each day. At night, they travel through trees in slow motion, feeding on leaves, shoots, and fruit. They rarely drink water, getting most of the moisture they need by eating leaves and licking dew.

Sloths are almost helpless on the ground, though they swim well. They are related to the armadillo and anteater.

Physical Description	Habitat and Range	Food	Life Span	Reproduction
20-25 in. long 9-20 lbs.	forests in parts of Central and South America	leaves, shoots, and fruit	as long as 30 years in captivity	1 young after a pregnancy of about 6 months

It's Classified Information
A Sorting Activity

This Project

In this project your students will sort and classify items such as leaves, rocks, or buttons. They will determine their own sorting criteria and draw a chart to illustrate it. Emphasis should be placed on how to develop and illustrate criteria.

Note: If you use the template (*Sorting*) provided on the CD-ROM, direct your students to follow the prompts to fill in the requested information.

Computer Skills

- word processing
- use of drawing and flow chart tools
- copying and pasting graphics
- using text boxes

Before Beginning

- Students should sort a group of tangible items and develop their own classification criteria.
- They should complete the planning sheet on Page 62.

Quick Steps

- Open a new document.
- Type the title of your document.
- Place an oval under the title.
- Type what was sorted in a text box inside the oval.
- Insert text boxes under the oval.
- Type the subgroups in the text boxes.
- Place arrows to point from the oval to the subgroups.
- Insert more text boxes and arrows under the subgroups.
- Save and print your document.

It's Classified Information
A Sorting Activity *(Cont.)*

Step 1   Open a new *Word* document.

Step 2   Click on the **Center** button on the **Formatting Toolbar**.

Set the font size to **24** or **36** points.

Type the title of your document.

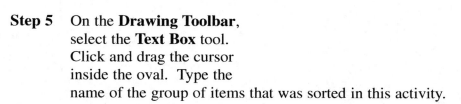

Step 3   From the **VIEW** menu, choose **Toolbars**, then **Drawing** if it is not already selected.

Step 4   Select the **Oval** on the **Drawing Toolbar**.

Click and drag the cursor to place the oval under the title.

To fill this oval with a color. click to the right of the **Fill Color** tool on the **Drawing Toolbar**. Select a color in the pop-up window.

Step 5   On the **Drawing Toolbar**, select the **Text Box** tool. Click and drag the cursor inside the oval. Type the name of the group of items that was sorted in this activity.

Step 6   On the **Drawing Toolbar**, select the **Arrow** tool. Click and drag to insert the arrow under the oval.

Step 7   On the **Drawing Toolbar**, click on the **Text Box** tool. Click under the arrow to place the text box. Type a subgroup in this text box.

Step 8   Insert more text boxes to the left and to the right of this text box. Type a different subgroup in each of the text boxes.

Step 9   Finish inserting arrows to point from the oval to each text box.

Step 10   If there are more characteristics under each subgroup, insert more arrows and text boxes.

Step 11   Save and print your document.

It's Classified Information A Sorting Activity (Cont.)

Planning Sheet

The title of your document: _____

Select a group of items to be classified or sorted: _____

Separate the items into subgroups by their characteristics. List those characteristics below. Keep in mind what the smaller groups have in common and how they are different from one another.

Describe those characteristics in the boxes below. Add text boxes and arrows as needed.

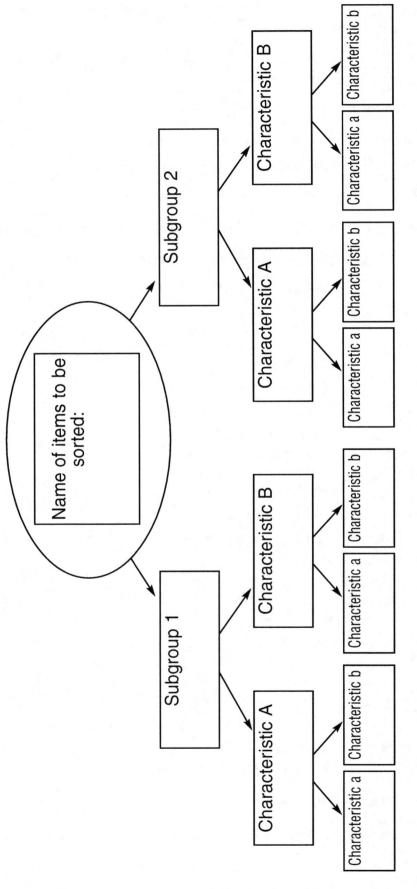

© *Teacher Created Resources, Inc.*

It's Classified Information A Sorting Activity (Cont.)

Sample

Classifying Buttons

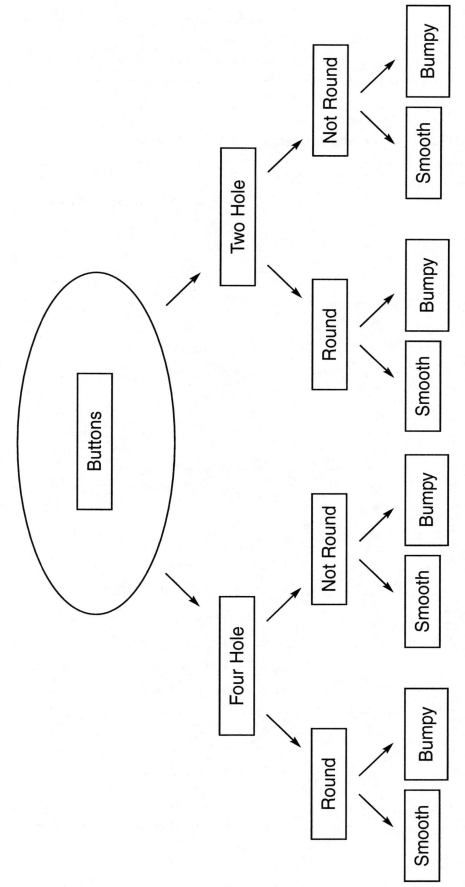

Testing, Testing: 1, 2, 3
Science Experiment

This Project

In this project your students will write an experiment they conducted or a hypothesis they tested using a scientific method.

Computer Skills

- word processing
- using the **Numbering and Bullet** tool
- using font style

Before Beginning

- Students should conduct an experiment or test a hypothesis.
- They should complete the planning sheet on Page 66.

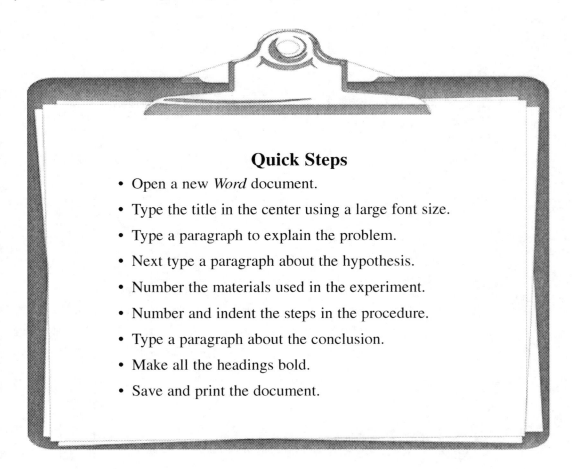

Quick Steps

- Open a new *Word* document.
- Type the title in the center using a large font size.
- Type a paragraph to explain the problem.
- Next type a paragraph about the hypothesis.
- Number the materials used in the experiment.
- Number and indent the steps in the procedure.
- Type a paragraph about the conclusion.
- Make all the headings bold.
- Save and print the document.

Testing, Testing: 1, 2, 3
Science Experiment *(Cont.)*

Step 1   Open a new **Word** document.

Step 2   Click on the **Center** button on the **Formatting Toolbar**.
Set the font size to **24** or **36** points.
Type the title of your report.

Step 3   Click on the **Align Left** button.
Reset the font size to **12** points.
Type "**Problem**".
Press the **Enter** or **Return** key. Type the problem. Press the **Enter** or **Return** key twice.

Step 4   Type "**Hypothesis**". Press the **Enter** or **Return** key. Type the hypothesis. Press the **Enter** or **Return** key twice.

Step 5   Type "**Materials**". Press the **Enter** or **Return** key. Type a list of all the materials used in the experiment. Press the **Enter** or **Return** key at the end of each item.

Step 6   Highlight the list of materials. From the **FORMAT** menu, select **Bullets and Numbering**. In the pop-up window, click the **Numbered** tab. Choose a style. Click **OK**. Press the **Enter** or **Return** key twice.

Step 7   Type "**Procedure**". Press the **Enter** or **Return** key. Type a list of all the steps in the procedure. Press the **Enter** or **Return** key at the end of each step.

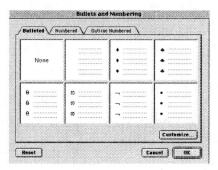

Step 8   Highlight the list of steps. From the **FORMAT** menu, select **Bullets and Numbering**. In the pop-up window, click the **Bulleted** tab or the **Numbered** tab. Choose a style. Click **OK**. Press the **Enter** or **Return** key twice.

Step 9   Type "**Conclusion**". Press the **Enter** or **Return** key. Type your conclusion.

Step 10   Highlight each of the headings by clicking on the word (**Problem, Hypothesis, Materials, Procedure,** and **Conclusion**). Click on the **Bold** button on the **Formatting Toolbar**.

Step 11   Save and print the document.

Testing, Testing: 1, 2, 3
Science Experiment *(Cont.)*

Planning Sheet

Problem: What did you want to find out?

_____

_____

Hypothesis: What was your hypothesis? What did you think you would find out?

_____

_____

Materials: What materials did you need?

1. _____

2. _____

3. _____

4. _____

Procedures: What steps did you follow?

1. _____

2. _____

3. _____

4. _____

Conclusion: What did you learn?

_____

_____

_____

_____

Testing, Testing: 1, 2, 3
Science Experiment *(Cont.)*

Sample

Bubbles Experiment

Problem

I wanted to find out which solution would work better for blowing bubbles—a solution with glycerin or one without glycerin?

Hypothesis

My hypothesis was that the solution with glycerin would blow better bubbles because the glycerin is stickier than soap and acts like glue.

Materials

- 8 ounces (225g) dish soap
- 32 ounces (900g) water
- 2 bowls (16 oz./450g water and 4 oz./112.5g soap in each)
- Wire or pipe cleaners
- Glycerin (2-4 oz.)

Procedure

- First, I made blowing tools with the wire and used the solution without glycerin. It made bubbles, but they broke quickly.
- Next, I used the solution with the glycerin and the same wire tools. I found that this solution worked better. The bubbles were bigger and lasted longer.

Conclusion

My hypothesis was correct. The solution with glycerin worked best. Bubbles lasted longer. The glycerin held the water and soap together.

What Goes Around...
Cycles and Processes

This Project

In this project your students will illustrate and write about a scientific cycle or process.

Note: You must have *Word 98/Office 97* or higher to use the template (*Cycle*) for this lesson. Direct your students to follow the prompts to fill in the requested information.

Computer Skills

- word processing
- using text boxes
- inserting, rotating and flipping block arrows

Before Beginning

Students should learn about a scientific cycle or process and complete the planning sheet on Page 71.

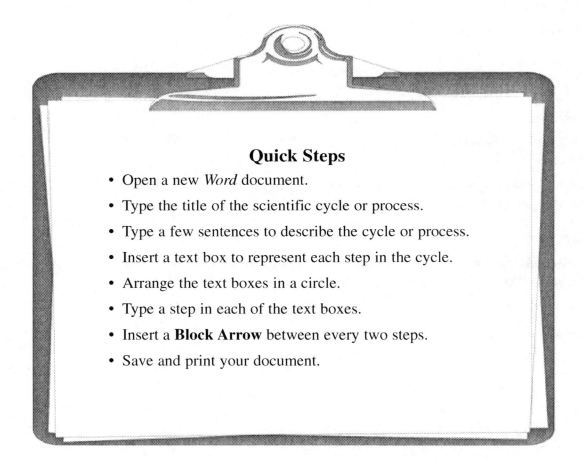

Quick Steps

- Open a new *Word* document.
- Type the title of the scientific cycle or process.
- Type a few sentences to describe the cycle or process.
- Insert a text box to represent each step in the cycle.
- Arrange the text boxes in a circle.
- Type a step in each of the text boxes.
- Insert a **Block Arrow** between every two steps.
- Save and print your document.

What Goes Around...
Cycles and Processes *(Cont.)*

Step 1 Open a new **Word** document.

Step 2 Click on the **Center** button on the
Formatting Toolbar.

Set the font size to **24** or **36** points.

Type the title or name of the process or cycle.

Reset the font size to **12** points.

Press the **Enter** or **Return** key twice.

Step 3 Click on the **Left** button.
Type a few sentences to describe the cycle or process.
Press the **Enter** or **Return** key twice.

Step 4 From the **INSERT** menu, choose **Text Box**. Click under the paragraph to place a text box.

Step 5 Repeat **Step 4** to place a text box to represent each of the
steps.
Arrange these text boxes in a circle.

Step 6 Type the steps in the text boxes in the clockwise order.

Step 7 From the **INSERT** menu, select **Picture**, then **AutoShapes** from the drop down menu.

Step 8 In the **AutoShapes** window, choose the third option,
Block Arrows. Select an arrow in the pop up window.

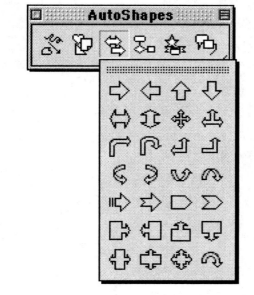

What Goes Around...
Cycles and Processes *(Cont.)*

Step 9 Click between the first and second steps to place the arrow. Size the arrow by dragging the handles around it. Move it by clicking inside and dragging it.

Step 10 From the **VIEW** menu, choose **Toolbars**, then **Drawing** from the drop down menu if it is not already selected.

Step 11 To change the direction of the arrow, click on **Draw** on the **Drawing** toolbar. Select **Rotate or Flip**, then **Free Rotate** in the pop up window.

The handles around the arrow will turn into little green circles. Drag one of the handles to rotate the arrow.

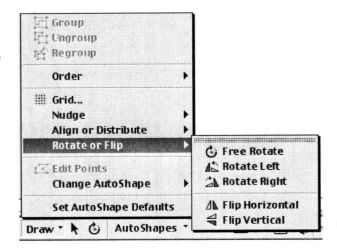

Step 12 To flip an arrow, click on **Draw** on the **Drawing Toolbar**. Select **Rotate or Flip**, then **Flip Horizontal** or **Flip Vertical** in the pop up window.

Step 13 Finish placing arrows between each two steps in the cycle or process.

Step 14 Save and print your document.

What Goes Around...
Cycles and Processes *(Cont.)*

Planning Sheet

Select a process or cycle such as photosynthesis, the water cycle, or a butterfly's life cycle. List the steps of this process in numerical order.

1. _____

2. _____

3. _____

4. _____

5. _____

6. _____

In the space below, sketch how you will illustrate the cycle or process.

What Goes Around...
Cycles and Processes *(Cont.)*

Sample

The Water Cycle

First there is precipitation or rain. The rain runs off the ground's surface. Then it evaporates. After it evaporates, it condenses. The condensation causes it to rain again. The cycle continues.

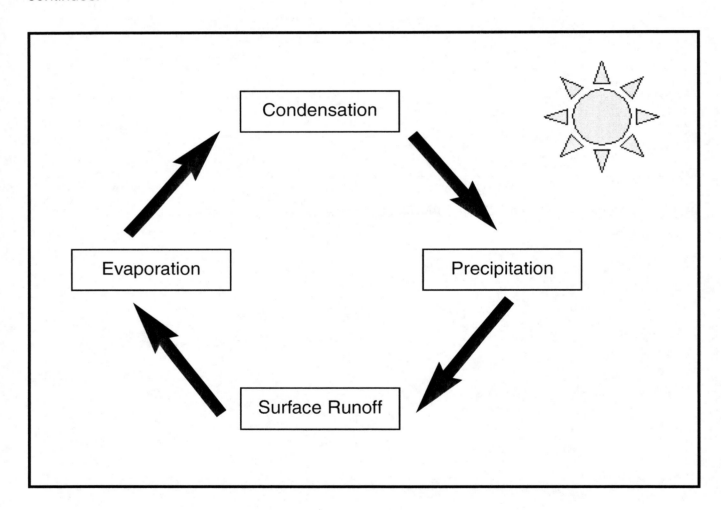

What's Next?
Pattern Page

This Project

In this project your students will use shapes to create patterns.

Note: You must have *Word 98/Office 97* or higher to do this lesson.

Computer Skills

- word processing
- inserting shapes
- copying and pasting shapes
- resizing and moving shapes
- filling shapes with colors

Before Beginning

Students should review patterns and complete the planning sheet on Page 75.

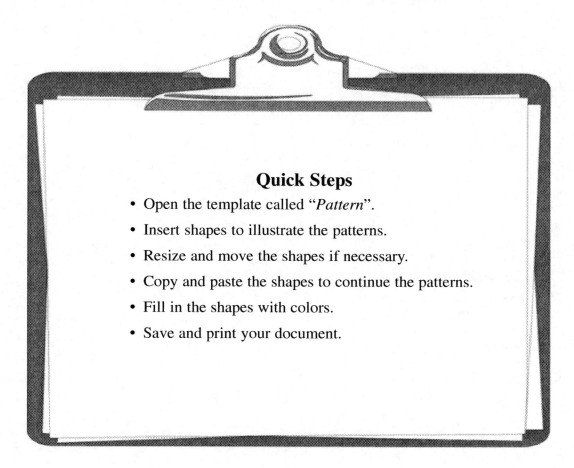

Quick Steps

- Open the template called "*Pattern*".
- Insert shapes to illustrate the patterns.
- Resize and move the shapes if necessary.
- Copy and paste the shapes to continue the patterns.
- Fill in the shapes with colors.
- Save and print your document.

What's Next?
Pattern Page *(Cont.)*

Step 1 Open *Microsoft Word*. From the **FILE** menu, select Open.

Step 2 Find the file called *"Pattern"*. Double-click on it to open the file.

Step 3 From the **INSERT** menu, choose **Picture**, then **AutoShapes** from the drop-down menu.

Step 4 From **AutoShapes**, select the second option, **Basic Shapes**. From the pop-up window, choose a shape. (**Note: Stars and Banners** or **Block Arrows** from **AutoShapes** also work well.)

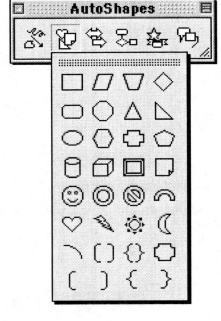

Step 5 Click under the first pattern to insert the shape. Resize this shape by dragging the handles around it.

Step 6 Move the shape by clicking inside and dragging it if necessary.

Step 7 Insert more shapes to continue the pattern.

Step 8 To use a shape more than once, select the shape by clicking on it. From the **EDIT** menu, select **Copy**, then **Paste**. A copy of this shape will appear on the top of the original shape. Move this copy by clicking inside and dragging it.

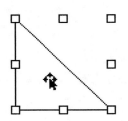

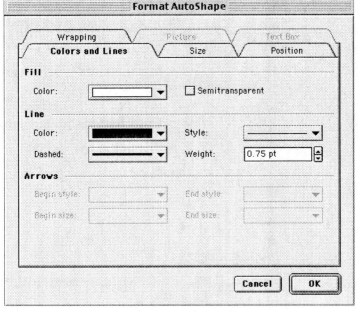

Step 9 To fill a shape with a color, double-click on the shape. In the **Format AutoShape** window, click the **Colors and Lines** tab. Select a new fill color. Click **OK**.

Step 10 Finish illustrating all the patterns.

Step 11 Save and print your document.

What's Next?
Pattern Page *(Cont.)*

Planning Sheet/Template

Sketch the following patterns:

A   A   A   B   A   A   A   B   A   A   A   B

A   B   A   A   A   B   A   A

A   B   C   C   A   B   C   C

What's Next? Pattern Page *(Cont.)*

Answer Sheet

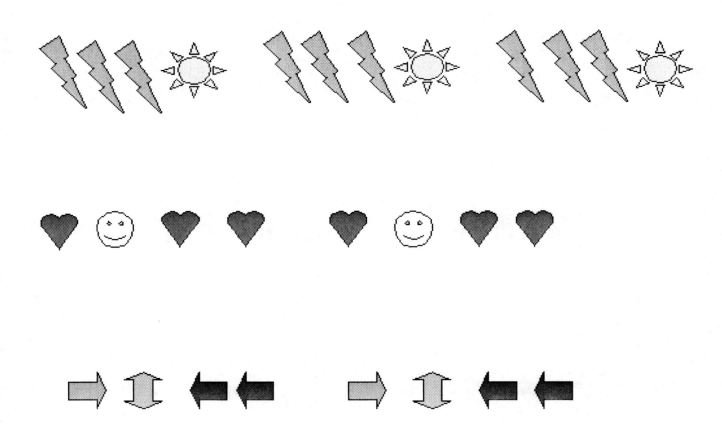

Playful Polygons
Pentagons, Hexagons and More

This Project

In this project your students will insert polygons such as triangles, rectangles, pentagons, hexagons, and octagons to fill in the boxes in a template.

Note: You must have *Word 98/Office 97* or higher to do this lesson.

Computer Skills

- word processing
- inserting shapes
- sizing and moving shapes

Before Beginning

- Students should review these polygons: right triangle, parallelogram, isosceles triangle, pentagon, trapezoid, hexagon, rectangle and octagon.
- Students should complete the planning sheet on Page 79.

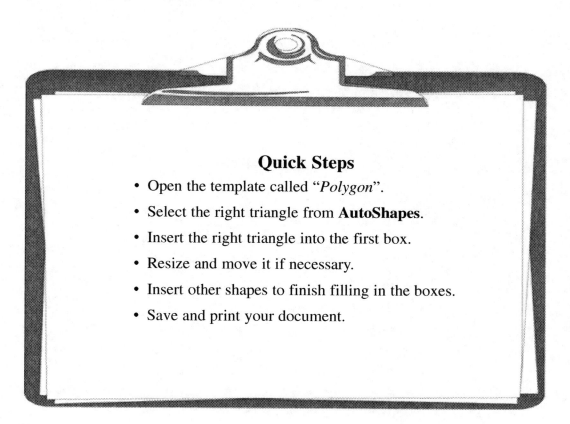

Quick Steps

- Open the template called "*Polygon*".
- Select the right triangle from **AutoShapes**.
- Insert the right triangle into the first box.
- Resize and move it if necessary.
- Insert other shapes to finish filling in the boxes.
- Save and print your document.

Playful Polygons
Pentagons, Hexagons and More*(Cont.)*

Step 1   Open *Microsoft Word*. From the **FILE** menu, select **Open**.

Step 2   Find the file called "*Polygon*". Double-click on it to open the file.

Step 3   From the **INSERT** menu, choose **Picture**, then **AutoShapes** from the drop-down menu.

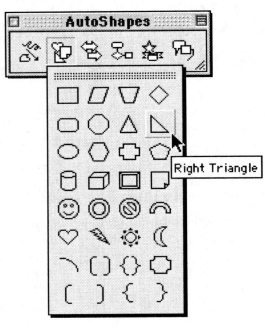

Step 4   From **AutoShapes**, select the second option, **Basic Shapes**. Choose the **Right Triangle** in the pop-up window.

Step 5   Click in the first box of the template where it says "**Right Triangle**" to insert the shape.

Resize this shape if necessary by dragging the handles around it.

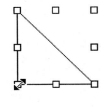

To move the shape, click inside and drag it.

Step 6   Finish inserting the correct shapes into the template.

Step 7   Save and print your work.

Playful Polygons
Pentagons, Hexagons and More*(Cont.)*

Planning Sheet

Sketch the following polygons:

Right triangle	Parallelogram
Isosceles triangle	**Pentagon**
Trapezoid	**Hexagon**
Rectangle	**Octagon**

Playful Polygons
Pentagons, Hexagons and More *(Cont.)*

Sample

Illustrate the following polygons:

Right triangle	Parallelogram
Isosceles triangle	**Pentagon**
Trapezoid	**Hexagon**
Rectangle	**Octagon**

All in a Day's Fun
Calculating Costs Activity

This Project

In this project your students will plan activities for a fun day and calculate its cost.

Note: If you use the template (*FunDay*) provided on the CD-ROM, direct your students to follow the prompts to fill in the requested information.

Computer Skills

- word processing
- inserting a table
- adjusting column widths
- calculating with the table formula function

Before Beginning

- Students should plan activities for a fun day, real or imaginary. What would they like to do or buy? How much would each activity or item cost?
- Students should complete the planning sheet on Page 83.

Quick Steps

- Open a new *Word* document.
- Type the title of your document.
- Type a description of what you would like to do for a fun day.
- Insert a table to itemize the activities.
- List each activity or item in the left column.
- List the cost of each activity or item in the right column.
- Use the Formula function to calculate the total costs.
- Save and print the document.

All in a Day's Fun
Calculating Costs Activity *(Cont.)*

Step 1   Open a new *Word* document.

Step 2   Click on the **Center** button on the **Formatting Toolbar**.

Set the font size to **24** or **26** points.

Type the title of your fun day.

Step 3   Reset the font size to **12** or **14** points.

Press the **Enter** or **Return** key twice.

Click on the **Align Left** button.

Type a paragraph describing the planned activities for your fun day. Press the **Enter** or **Return** key twice.

Step 4   From the **VIEW** menu, select **Page Layout** if it is not already selected.

Step 5   From the **TABLE** menu, select **Insert Table**. In the dialog box that pops up, type in **2** for the number of columns; for the number of rows, type in the number from your planning sheet. Select **Auto** for the column width. Click **OK**.

Step 6   Click in the first cell in the left column. Type one item or activity. Click in the first cell in right column. Type the cost of this item or activity.

Step 7   If you need to adjust a column width, move the cursor over the vertical line of the column. When the cursor changes to a resizing arrow, click and drag the line to the left or right.

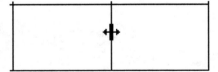

Step 8   Finish filling in the items or activities in the left column and their costs in the right column.

Step 9   In the last cell in the left column, type "**Total Costs**".

Step 10   Click in the last cell in the right column. From the **TABLE** menu, select **Formula**. In the dialogue box that pops up, "**=SUM(ABOVE)**" will appear by default. Click **OK**.

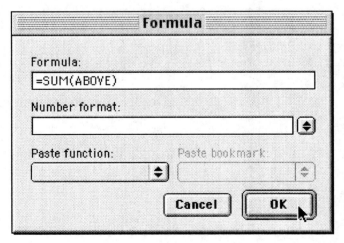

Step 11   Save and print your document.

All in a Day's Fun
Calculating Costs Activity *(Cont.)*

Planning Sheet

Decide on some recreational activities for a day. Write a few sentences to describe the activities. What would you like to do and buy?

_____

_____

_____

_____

In the left column, list what you would like to do and buy. In the right column, list the cost for each item or activity.

Total number of items and activities: _____

Number of rows for the table (the above Total + 1): _____

All in a Day's Fun
Calculating Costs Activity *(Cont.)*

Sample

A Day at an Amusement Park

I would like to go to the amusement park. At the park, I will ride the roller coaster. I will eat a corn dog and popcorn. I will play a tossing game.

Park admission	$5
Roller coaster	$1.50
Corn dog	$2
Popcorn	$0.75
Tossing game	$1.00
Total costs	$10.25

It's All in How You Cut It
Fraction Slices

This Project

In this project your students will divide figures into fractions.

Note: If you use the template (CutIt) provided on the CD-ROM, direct your students to follow the prompts to finish the activity.

Computer Skills

- word processing
- using the **Line** tool

Before Beginning

- Students should understand the concept of fractions.
- Students should complete the planning sheet on Page 87 in pencil.

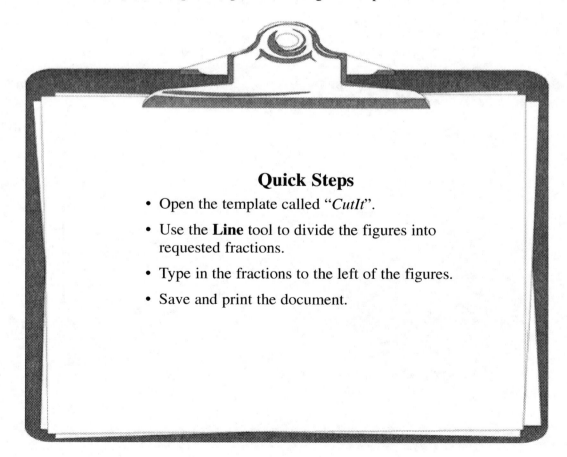

Quick Steps

- Open the template called "*CutIt*".
- Use the **Line** tool to divide the figures into requested fractions.
- Type in the fractions to the left of the figures.
- Save and print the document.

It's All in How You Cut It
Fraction Slices *(Cont.)*

Step 1   Open *Microsoft Word*. From the **FILE** menu, select **Open**.

Step 2   Find the file called "**CutIt**". Double-click on it to open the file.

Step 3   From the **INSERT** menu, choose **Pictures**, then **AutoShapes** from the drop-down menu.

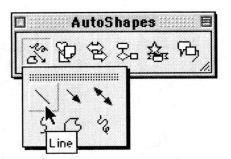

Step 4   In the **AutoShapes** window, click on the first option, the **Lines** tool. Select the **Line** in the pop-up window.

Step 5   Inside the first figure, click and drag the cursor to draw lines to slice the figure into requested fractions.

Step 6   Click above the line to the left of the figure. Type the fractions.

Step 7   Repeat **Steps 4 to 6** to finish cutting the figures into requested fractions and to type the fractions on the lines.

Step 8   Save and print the document.

It's All in How You Cut It
Fraction Slices *(Cont.)*

Planning Sheet

With your pencil, draw straight lines to divide each figure into requested fractions. Type the fraction on the line provided to the left of each figure.

Use the slash key / to separate the numerator from the denominator as in this example: 2/2

1. Divide the rectangle into fourths.	2. Divide the circle into eighths.
_____ 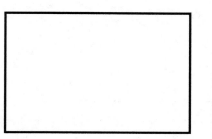	_____ 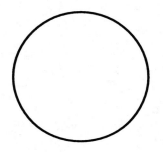
3. Divide the triangle into halves.	4. Divide the hexagon into sixths.
_____ 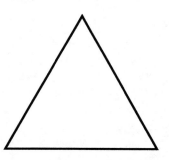	_____ 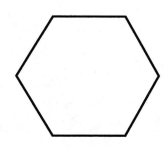
5. Divide the octagon into eighths.	6. Divide this shape into sixths.
_____ 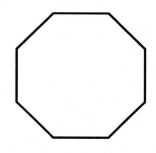	_____ 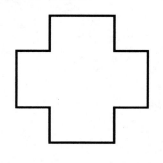

It's All in How You Cut It
Fraction Slices *(Cont.)*

Answer Sheet

Use the Line tool to divide each figure into requested fractions. Type the fraction on the line provided to the left of each figure.

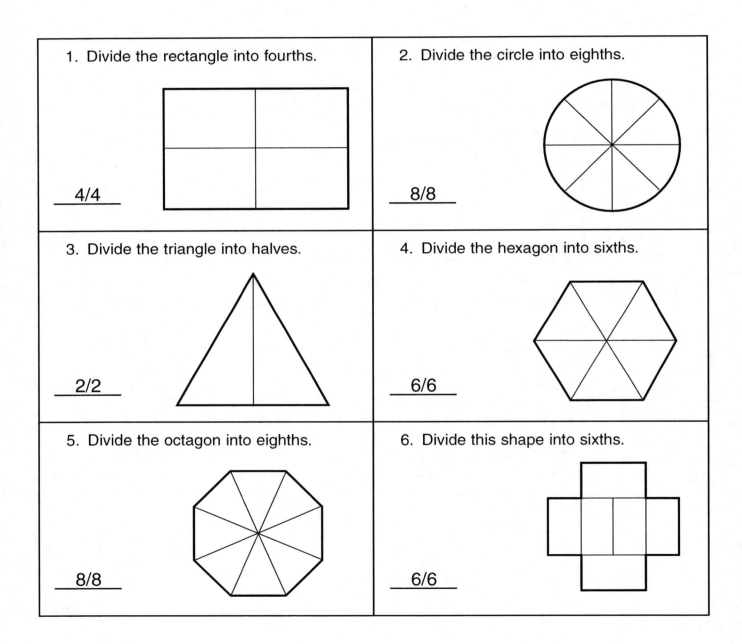

1. Divide the rectangle into fourths.

   4/4

2. Divide the circle into eighths.

   8/8

3. Divide the triangle into halves.

   2/2

4. Divide the hexagon into sixths.

   6/6

5. Divide the octagon into eighths.

   8/8

6. Divide this shape into sixths.

   6/6

What's Your Angle?
Acute and Obtuse Triangle Assignment

This Project

In this project your students will use *Microsoft Word* to transform an isosceles triangle to an acute triangle, an obtuse triangle, and a right triangle.

Note: You must have *Word 98/Office 97* or higher to do this lesson.

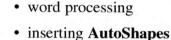

Computer Skills

- word processing
- inserting **AutoShapes**
- copying and pasting **AutoShapes**
- transforming and resizing **AutoShapes**
- inserting text boxes
- resizing and moving text boxes

Before Beginning

- Students should review acute, obtuse, and right angles.
- They should complete the planning sheet on Page 91.

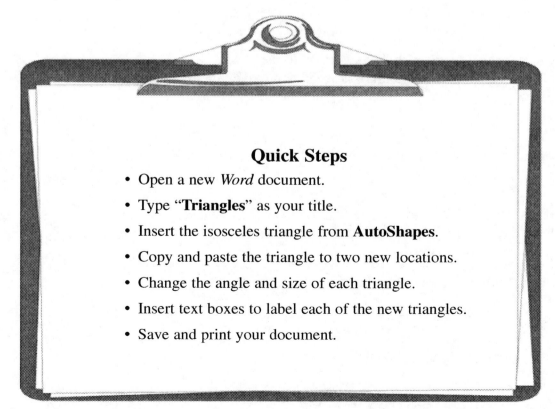

Quick Steps

- Open a new *Word* document.
- Type "**Triangles**" as your title.
- Insert the isosceles triangle from **AutoShapes**.
- Copy and paste the triangle to two new locations.
- Change the angle and size of each triangle.
- Insert text boxes to label each of the new triangles.
- Save and print your document.

What's Your Angle? Acute and Obtuse Triangle Assignment *(Cont.)*

Step 1   Open a new *Word* document.

Step 2   Click on the **Center** button on the **Formatting Toolbar**.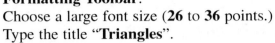
Choose a large font size (**26** to **36** points.)
Type the title "**Triangles**".

Step 3   From the **INSERT** menu, choose **Pictures**, then **AutoShapes** from the drop-down menu.

Step 4   In the **AutoShapes** window, choose the second option, **Basic Shapes**. Select the isosceles triangle in the pop-up window.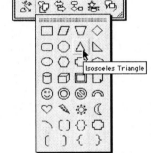

Step 5   Click under the title to insert the triangle.

Step 6   With the triangle still selected, pull down the **EDIT** menu. Select **Copy**, then **Paste**. A copy of this triangle will appear on top of the original. Drag the copy to a new location under the first triangle, leaving some room in between.

Step 7   From the **EDIT** menu, select **Paste** again. A new copy of the first triangle will appear. Drag this copy to a new location under the second triangle, also leaving some room in between.

Step 8   Click on the first triangle.
Drag one of the handles till you
get an acute triangle.

Step 9   Click on the second triangle. Drag one of the handles till you get an obtuse triangle. If necessary, move this triangle by clicking inside and dragging it.

Step 10   Click on the third triangle. Drag the yellow handle till you get a right triangle.

Step 11   From the **INSERT** menu, select the **Text Box**. Click under the first triangle to insert the text box. Type "**Acute Triangle**". Size and move the text box to line up with the triangle.

Step 12   Insert another text box under the second triangle. Type "**Obtuse Triangle**". Size and move the text box to line up with the triangle.

Step 13   Insert another text box under the third triangle. Type "**Right Triangle**". Size and move the text box to line up with the triangle.

Step 14   Save and print your document.

What's Your Angle?
Acute and Obtuse Triangle *(Cont.)*

Planning Sheet

Sketch the following triangles.

Acute Triangle

Obtuse Triangle

Right Triangle

Sample

Triangles

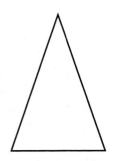

Acute Triangle

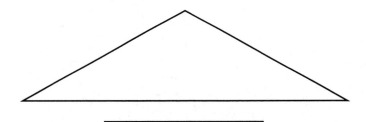

Obtuse Triangle

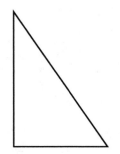

Right Triangle

Glossary

Align Left—The text is arranged to form a straight left margin.

Align Center—The text is placed at the center of each line.

Align Right—The text is arranged to forma straight right margin.

AutoSum—Addition of a column of numbers in the Table feature.

Cell—An area in a table where data may be entered.

Clip art—Pre-drawn and preformatted graphic that can be placed in a document.

Copy—The command that copies an item onto the Clipboard to be saved for later use elsewhere.

Dialogue Box—This window appears when a selection has been made. It will ask you questions about how you want the computer to proceed in the activity that has been chosen.

Fill—The color or pattern used in solid areas of a shape.

Font— the design style of the typeface.

Handles—The small boxes that appear when sizing and moving graphics.

HTML—Hypertext Media Language used on the Internet for coding web pages.

Import—To bring in graphics or text that were not created in *Microsoft Word*.

Indentation—The start of a line that is further from the margin than the remainder of the Document.

Landscape—The document is turned so that the long side of the document is the top edge of the page.

Point Size—The size of the letters that are typed.

Text Box—A box inserted in a document that allows typing outside of the normal document format.

Toolbar—The pallet of tools that is available for use.

Appendix: Tips and Tricks

Adding Borders or Shading to a Document

To add a border or shading to a document, type your text first, then from the **Drawing Toolbar**, select the **Rectangle** tool. Draw a rectangle to frame your document. The text will be covered. From **Draw** on the **Drawing Toolbar**, select **Order**, then **Send Behind Text**. The text will appear inside the frame. While the frame is highlighted, select the **Shadow Tool** from the **Drawing Toolbar**. You will be presented with many shading options. Experiment until you find the one you want to select.

Adding a Picture

Use Clip Art

Many companies sell prepared pictures called clip art. These can be placed in any *Microsoft Word* document. If you have clip art available to you, this is how you can add it. From the **Insert menu**, go down to **Picture**. Move the mouse to the right and select **From File**. Then locate the file from the proper place on your computer's hard drive, CD-ROM, Zip or floppy disk. If you select **Clip Art**, the **Microsoft Clip Art Gallery** will pop up. But if you are using a different Clip Art Library, you need to click on the **Import** button.

Use Other Programs

Look in an electronic encyclopedia or other electronic reference source for a picture that meets your needs. Select the picture. Copy it to the computer's clipboard by pulling down the **EDIT** menu and selecting **Copy**. Then click on the location in your document where you want it to be placed. From the **EDIT** menu, select **Paste**.

Use a Scanner

If you have a scanner, you can use a photograph or drawing to make a digitized file to import into your document. Save the scanned image according to the directions for your scanner. Then you can insert the image from the **INSERT** menu by selecting **Picture**, then **From File**.

From the Internet

You can easily copy pictures from web pages. With both your browser program (*Netscape, Internet Explorer*, etc.) and *Microsoft Word* open, click on the picture you want to select while viewing the web page on your browser. Select **Copy Image**. Then switch to the *Microsoft Word* document. Click on your document where you want to place the picture. From the **EDIT** menu, select **Paste**.

Comments

The Commenting feature allows you to write visible notes to yourself or others who will review your work that are not a part of the actual document. To write a comment, from the **VIEW** menu, select **Toolbars**, then **Reviewing**. The first icon on the reviewing toolbar is the **Insert Comment** tool.

Creating Stationery

Stationery is a way to save a file where, when it's opened, a new file is created with the text that was previously entered. When you save the new file, the original remains the same. This feature is particularly useful at a classroom computer center where you have an assignment prepared for your students. They will be able to open the file from the desktop, or wherever you save it, yet they will not "accidentally" override your original document when they save their work.

To create stationery, when saving your document, go to **Save As** from the **File Menu**. On the bottom of the dialog box that appears, it will say, "**Save File as Type**" with **Word Document** next to it. Click on the arrow next to **Word Document**, and scroll down to **Stationery**. Select **Stationery**. Then save your document.

Customizing The Toolbar

If there are some features in *Microsoft Word* that you use more often than others, you might want to customize your toolbars. From the **VIEW** menu, select **Customize**. You will be able to select all of your favorite features and deselect those you don't use.

Getting Help

When you need help remembering how to complete a task or need to know what task you need to do, you can consult *Microsoft Word* Help. From the **HELP** menu, select *Microsoft Word* **Help**. A window will pop up in which you can type in your question. *Microsoft Word* Help will locate the answer.

For a list of general and specific help topics, from the **HELP** menu, select **Contents and Index**. The **Contents** section will list general help topics. Click on the **Index** tab for a list of very specific topics.

Locating a Specific Word

If you are using a *Macintosh* operating system, use the **Command-F** keystroke (the key with the apple next to the space bar and the F key). While the apple key is pressed, click on the **F-key**. A **Find and Replace** dialogue box will pop up. Type in the word you want to locate. You will be able to locate any and all occurrences of that word.

If you are using a **PC** operating system, use the **Control-F** keystroke.

Replacing a Specific Word

To replace a specific word, follow the instructions above for locating a word. Then in the **Find and Replace** dialogue box, click on the **Replace** tab. You will be able to replace any and all instances of the word as desired.

Saving a Document for Use on Other Computers

Saving for different versions of *Microsoft Word*

Perhaps you are working on a document both at home and at school on different computers with different versions of *Microsoft Word*. *Microsoft Word* enables you to save documents in older versions of *Word*. For example, if you have *Word 98* at home, but *Word 5* at school, you can save the documents you do at home in *Word 5*, so that they can be opened and read at school. (You may lose some formatting, but the bulk of your work will remain in tact.)

From the **FILE** menu, go to **Save As**. On the bottom of the dialog box that appears, you will have many options to **Save File Type As...** Select the version of *Word* that you need. Remember to save in the oldest version in which you intend to open the document.

Saving on a Mac to use on a PC

If you work on both Macs and PCs, save your work on a PC floppy or Zip disk. Macs can use PC disks, though PC can't read Mac disks.

Saving for use in other word processing applications

If you plan to open your *Word* Document in another application, you can save it in **Text Only**, though font style and size formatting will be lost. You can also save it in **Rich Text** formatting without loss of formatting.

Word Wrapping

Word Wrapping or **Text Wrapping** allows you to wrap words around shapes. You can select the type of wrap to use in a certain situation by first selecting the shape and then selecting the **Text Wrapping** tool from the **Picture Toolbar**. You can also choose text wrapping for simple shapes from the **AutoShapes** menu under **Format**. In the **AutoShapes** dialog box, you will have options on which wrapping style to use and where you want to place the text.

Use the **Text Wrapping** tool to give your projects a more professional layout and appearance and to maximize space usage, particularly in table cells or documents of two or more columns.